FIELD PROPULSION

BY CONTROL OF GRAVITY

THEORY AND EXPERIMENTS

STOYAN SARG, Ph.D.

Copyright © 2009 by Stoyan Sarg. All rights reserved.

Fair Use Notice: The book contains a few figures, images and text citations from peer-reviewed journals, books and public domains the use of which has not always been specifically authorized by the copyright owner. Such material with proper citation and acknowledgement is included for the purpose of scientific research, comments and criticism in an effort to advance our understanding of Nature and the Universe. We believe this constitutes a 'fair dealing' and 'fair use' of any such copyrighted material as provided for in Part III Section 29 of the Canada Copyright Law and section 107 of the US Copyright law.

Images on the book covers:
- Front cover: Figures 8 and 9 are from the author's patent application "Method and apparatus for spacecraft propulsion with a field shield protection"
- Back cover:
  - Left column: Vacuum cell and optical signature of activated plasma in partial vacuum for 3 different gases
  - Right column: Picture of the author and one of experiments demonstrating the gravito-inertial SARG effect.

CONTENTS

List of abbreviations... V
Disclaimer... VI
Acknowledgements.. VII

CHAPTER 1. The ory of alternaticve space consept and propulsion mechanism of new kind..5
1.1 The importance of space concept. Historical overview...........5
1.2. Brief summary of the BSM - Supergravitation Unified Theory (BSM-SG)..10
1.3. Considerations about matter, space, energy and gravitation according to the BSM-SG theory 11
1.4. Gravitational and inertial mass..13
1.5. Longitudinal waves..16
1.6. On possibility of changing the mass of a solid object according to BSM-SG theory...18
1.7. Change of gravity by disturbance of the CL space surrounding the object...22
1.8. Field Propulsion by unidirectional change of the Newtonian mass of a body..29
1.9. Hypothesis of Field Propulsion mechanism......................31
1.10. Physical effects accompanying the Field Propulsion mechanism..34
1.11. Signature of some characteristic features of the new propulsion mechanism in the observed UFO cases....................35
1.12. Field Propulsion side effects..37

CHAPTER 2. Heterodyne Resonance as a physical mechanism for invoking a Field Propulsion..41
2.1. Accessing the Compton frequency by using the oscillation properties of electron...41
2.2. Quantum mechanical spin of the electron........................49
2.3. Heterodyne Resonance mechanism and SARG effect..........50

CHAPTER 3. Technical realization of the Heterodyne Resonance Mechanism. Gravito-inertial effect called a Stimulated Anomalous Reaction to Gravity (SARG)..55
3.1. Brief historical overview..55
3.2. Gravito-inertial phenomenon..56

3.3. Experiments for investigation of the Heterodyne Resonance Mechanism and demonstration of the SARG effect...................57
3.4. Conclusions:..74

CHAPTER 4. Method and apparatus for spacecraft propulsion with a field shield protection (Invention).................................75
Background of the invention..75
Summary of the invention..78
Brief description of drawings..79
Detailed description..80
I claim:..101
References (for this patent application):.......................103

CHAPTER 5. Field shield protection and Supercommunication: Hypothesis...109
5.1. Field shield protection: Hypothesis......................109
5.2 Unconventional method of communication by Longitudinal waves. Hypothesis..112
5.3. Wilbert Smith and his Tensor Coil..........................114
5.4. Some tests of the Smith Coil....................................119

SUMMARY AND CONCLUSIONS...............................123

APPENDIX 1 Table of Ether drift experiments..............127
APPENDIX 2 Physical model of the electron
          (article in Physics Essays)..........................129

About the author..159
References..163

LIST OF ABREVIATIONS

BSM-SG	Basic Structures of Matter –Supergravitation Unified Theory
SG	Super Gravitation
CL	Cosmic Lattice
ZPE	Zero Point Energy
ZPE-S	Static ZPE
ZPE-D	Dynamic ZPE
LWs	Longitudinal Waves
SPM	Spatial Precession Momentum (vector)
MQSPM	Magnetic Quasisphere (hodograph) of SPM
EQSPM	Electrical Quasisphere (hodograph) of SPM
SGSPM	Super Gravitation Spatial Precession Momentum (vector)
SARG	Stimulated Anomalous Reaction to Gravity effect
UFO	Unidentified Flight Object

DISCLAIMER

The Field Propulsion invoked by the Heterodyne Resonance Mechanism is a new field of research involving unknown so far physical effects. Experiments in this field require a lot of precautions. From one hand they involve the use of very high electrical fields in order of tens, hundred or even millions of volts. From the other hand the effect is accompanied with emission of Longitudinal (referenced also as Scalar) waves. They may cause problem not only for sensitive electrical equipment if not properly used but could be also potentially hazardous for living species and humans. Isotropic Longitudinal waves dissipate fast with the distance but may penetrate through shields and can propagate by ground connections. Presently such waves at proper doze are used for sterilization of products, while the physical origin of these waves has not been correctly understood [54]. Experiments related to the gravito-inertial SARG effect must be provided by highly qualified researchers and in special environments. Individual researchers must be aware that they provide experiments in this field at their own risk and responsibility. The author of this book cannot be held responsible for any injury, damage or loss of property that may eventually occur by improper experimental setup or test provided in not suitable environment.

Acknowledgements

First and foremost, I would like to thank God for giving me the inspiration and strength to complete this project. I wish to thank my family, my wife Denka and son Ivor, for their moral support during many years of extensive research and experiments. I am thankful to Patricia Klambauer for archiving and cataloguing my BSM-SG unified theory in the National Library of Canada. I appreciate the communications with Dr. Andrew Michrowski from the Planetary Association for Clean Energy, Canada, and his encouragement for using the derived theoretical models of BSM-SG theory to study the space energy and gravity. I am thankful to the editorial board of the Physics Essays journal and especially to Dr. Emilio Panarella for publishing my first article related to BSM unified theory. I would like to express my special thanks to Acad. Dr. Asparuh Petrakiev for the organized seminars, summer conferences and useful discussions. I highly appreciate the help of Nikolaos Balaskas from the York University, Ontario, Canada with whom I frequently discussed my ideas. He supplied me with valuable material about UFO subject matter and encouraged my experimental work for solving the mystery of this phenomenon. I am thankful to Dr. Andrej Tenne-Sens for the organized seminar on my BSM-SG theory that led to publication of a review of my books in Physics in Canada journal. I appreciate the useful discussions with William Treurniet and his help in editing this book. I would like to thank the organizing committees of the Natural Philosophy Alliance, the Society for Scientific Exploration and the Integrity Research Institute for the opportunity to present my theoretical and experimental work. I also appreciate the useful discussions I had with Dr. Todor Proytchev, Dr. Rumen Kakanakov, George Hathaway, Tom Bearden, Penn Penev, Angel Manev and Dimitar Valev. My special thanks go to Ator Sarkisoff for the technical help in my experimental work. Finally I am thankful to my university colleagues and Internet group collaborators for the useful discussions and comments.

INTRODUCTION

By its very nature the Aether is a vibrating medium.
Isaac Newton, 1675

The focus of this book is on a new kind of propulsion system relying on physical principle that differs from the principle of jet propulsion systems. It falls into the category of the observed UFO phenomenon. More specifically, it is related to UFO objects initially known as flying saucers. In this respect, only real spacecraft are considered when mentioning UFO phenomenon in this book. This excludes any speculative suggestions that the UFO is a paranormal effect.

The nature of UFO phenomena is still considered to be enigmatic because the observed and registered physical effects are not explainable from the point of view of Modern Physics. During the past 50 years significant observational material has been accumulated by airplane pilots, military personal and civilians. While military observations have been highly classified during the period of the cold war, some archives are already open to public and scientific analysis. On 17 Aug 2009, the British government announced opening the archives of UFO files recorded up to 1990. Over 4,000 pages have been posted online documenting 800 alleged encounters during the 1980s and 1990s. Over the past three years the British Ministry of Defense gradually released previously secret UFO papers after facing Freedom of Information demands. Presently UFO observations are mostly supplied by individuals using digital imaging. At the same time, a large number of hoaxed images and videoclips were produced using computer image processing techniques. The experienced UFO researchers, however, may distinguish the hoaxes from real UFO images. The continuing accumulation of UFO observations forces us to raise a question? Does our understanding of fundamental laws in Physics correctly reflect the reality and where we must look for possible errors?

The author of this book devoted significant time and effort in investigating the history of Physics over the past 150 years. Special attention was focused to the critical time at the beginning of the 20th century. During this time, the fundamental postulates of Modern Physics were formulated. The author's main focus was on the concept of space (physical vacuum) since it plays a fundamental role in the formulation of the adopted postulates. In parallel with

Introduction

this he analyzed a large number of experiments from the point of view of an alternative concept of space. The analysis led to a conclusion that the accepted concept of space (physical vacuum) does not correspond to reality. A new concept of the physical vacuum was suggested leading to new physical models about space and elementary particles. The new concept and derived physical models appear to work excellently through all fields of Physics: - Particle physics, Quantum Mechanics, Newton's physics, Theory of Relativity and Cosmology. As a result, the author developed a treatise called Basic Structures of Matter – Supergravitation Unified Theory (BSM-SG) [1,2,3,4,5,6,7]. The most amazing feature of the alternative space concept is that it permits to go beyond the adopted postulates of Newton's laws of gravity and inertia, Einstein's Theory of Relativity and the rules of Quantum mechanics. Extended to Cosmology, it allows to unveil in what state could be the matter in the "black holes", currently known as the singularity problem. One of the most useful derivations with potential applications is the unveiled relation between electric and magnetic fields on one side, and gravity and inertia of a material object on the other.

In the field of UFO research, the author predominantly focused on the physical effects of observed and documented UFO cases. Enormous amounts of data were investigated from the new point of view of BSM-SG theory. The observed physical effects match quite well the theoretical predictions of BSM-SG theory about the possibility of modifying the gravitational and inertial mass of an object by proper modulation of the parameters of the physical vacuum. The resultant effect is a new kind of propulsion mechanism. Its principle is distinguished from known jet propulsion systems based on the principle of inertial reaction.

In order to help the reader to understand the physics and the propulsion mechanism of UFO, a brief extract-summary from the treatise BSM-SG is included in the Chapter 1 of this book. The BSM-SG theory shows that the Newtonian mass is not equal to matter, so it cannot be assumed as an unchanged constant. It depends on the state of space, known as the physical vacuum. Therefore, the Newtonian mass of a body can be modified (changed) by proper modulation of the parameters of space (physical vacuum). Such a change includes both the gravitational and inertial mass of the object. Furthermore, not only the mass of the material object can be changed but also the invoked change may appear as a vector, i. e., the change may affect the magnitude and a direction of the gravitational (inertial) mass. This means creation of a specific force field, which does not have a counterpart in known Newtonian physics. This is a Field Propulsion effect predicted in the BSM-SG theory (referred to as a manipulative displacement in BSM-SG, Chapter 13) and proved experimentally in the laboratory. It is called a "Stimulated Anomalous Reaction to Gravity" (SARG) effect. In fact, the SARG effect is a mass modulation (decrease or increase) able to create a unique force field. SARG is not an ordinary EM effect, so the Einstein equation $E = mc^2$ is not applicable to it (the change of mass does not convert to energy in this case). The possible advantages of this type of mass modulation are acceleration with reduced energy, the ability to make sharp turns (large accelerations) which is often observed in UFO cases, and a lack of turbulence because the mass of the surrounding atmospheric molecules is also affected. The technical realization of the SARG effect involves EM activated neutral plasma. A spacecraft based on the SARG effect does not need to carry propellant material since the invoked force field is not based on the known reactive principle. For this reason, the UFO propulsion mechanism is completely different from the jet propulsion used in rockets. In the past 15 years, significant research on Electrogravity propulsion has been done world wide for realization of effective massless propulsion. However, the results are still much below expectations. The main reason is a lack of fundamental physical understanding, so the research was based on empirical methods wrongly addressed by many researchers as "physics". In these reported experiments, the "physics" is strongly dependent on the particular design since the fundamental issue is not understood. Without such understanding, the research is time consuming, not efficient, and consequently extremely expensive.

    The purpose of this book is to provide answers to some long standing questions by unveiling the mystery of the UFO spacecraft. More specifically, the primary goal is to provide a physical explanation of the UFO propulsion mechanism, its enigmatic motion and the accompanying side effects. Even the well advanced UFO technology has side effects some of which may disturb radio communications and even be harmful to living organisms in close proximity. For this reason, even if the technology is acquired, the UFO spacecraft cannot be considered as a replacement for airplanes as some people might think. The described new propulsion mechanism is suitable only for interplanetary and interstellar trips. The secondary goal of this book is to suggest technical methods for realization of this propulsion mechanism. The third goal is to show that the interplanetary UFO spacecraft probably generate a protective field shield against micrometeorites. The fourth goal is to show that, for most UFO spacecraft, the known EM waves are not convenient for communications between the crew and the external world or other UFO because they are disturbed by three factors - the plasma around the spacecraft, the Faraday cage effect from the spacecraft body, and the time change inside the spacecraft. For these reasons other types of waves must be used. They could be generated by a different technique and detected by a different type of receiver. Ordinary EM receivers are not capable of detecting these waves.

Amongst the most serious UFO researchers providing useful analysis of the observed and registered physical effects are Paul Hill [30] and Peter Sturrock [31]. Some UFO researchers offer probable explanations, but they are not supported by serious scientific analysis. A large number of researchers in the Electrogravity field are focused on experiments but without deep physical understanding, so they don't associate their research with UFOs. Until now, however, no one other than the author of this book have expressed the idea that a new physics based on an alternative space concept is needed to understand the physical mechanism of the UFO's propulsion system. The BSM-SG theory was published in 2001 initially electronically and then as a book in 2006 [4], while in the same period it was reported at a number of international conferences [3,5,8,9]. The author first reported the principle of the envisioned massless propulsion mechanism in the International Conference on Future Energy (COFE-2006) 22-24 Sep 2006, Washington DC [8]. In 2007, the author reported and demonstrated a massless motion experiment at the 26 Annual meeting of Society of Scientific Exploration (SSE) [9]. In 2008 the author provided further lab experiments proving that the observed effect is a change in the gravito-inertial mass of the object under test. The newly discovered physical effect was called Stimulated Anomalous Reaction to Gravity (SARG) effect. It was reported the same year at the 27 Annual conference of SSE [10]. In the same year (2008) the author of this book submitted a patent application to the Canadian patent office for an invention entitled "Method and Apparatus for Spacecraft Propulsion with a Field Shield Protection" [11]. Extract from the patent application is included in Chapter 4.

CHAPTER 1. Theory of alternative space concept and propulsion mechanism of new kind.

This chapter contains a brief summary of the Basic Structures of Matter – Supergravitation unified theory, [1,4] and derived conclusions for new practical applications. Among the major potential applications is a new kind of propulsion mechanism.

1.1 The importance of space concept. Historical overview.

It is reasonable to consider that if space is not empty but containing some physical medium beyond our technological level of detection, the latter will be responsible not only for propagation of the Electrical and Magnetic fields but also for Newtonian gravity. Furthermore, it will also define the inertia due to equality between gravitational and inertial mass. The properties of this medium should define the known space-time properties of space. Then if we are able to modulate in some way the parameters of this space medium, we may expect to observe some change in the gravitational and inertial mass of an object, accompanied by some unusual effects of EM wave propagation.

The concept of space adopted in contemporary physics plays a crucial role in our understanding of Nature and the Universe. The space concept has been changed a few times in the past. It is reasonable to raise a question: Is the currently adopted space concept a final truth? Great discoveries in Physics were made even before the technological era of the 20th century. At that time the Ether concept of space was adopted. Famous names like Newton, Faraday, Ampere, Lord Kelvin, Maxwell, Nikola Tesla and many others are related to discoveries envisioned by the Ether concept of space.

The beginning of the 20th century marks a period of significant transition in our understanding of space. The previously dominant Ether concept was replaced by a new concept of the physical vacuum now known as a space-time continuum that incorporates the relativistic effects. The new concept was of fundamental importance for all further developments of some fields in Physics, particularly Theoretical Physics and the highly abstract branch of Mathematical Physics. At the same time, the adopted concept put a boundary on deeper understanding of some fundamental physical issues about fields. More specifically, it could not provide answers to many fundamental questions, for example: why is the speed of light constant; what defines inertia and relativistic effects; does gravity propagate instantaneously or with the speed of light, and many others.

The concept of space adopted at the beginning of the 20th century played a crucial role in the definition of a number of postulates on which Modern Physics relies. According to this concept, space is empty but light and electromagnetic waves travel with a very precise constant velocity. Furthermore, Einstein postulated that velocity of light is independent of the moving observer, and this is a corner stone for his theory of Special Relativity. In fact, Special Relativity adopted the previously known Lorenz formulae for motion in an Ether medium, but because the above mentioned postulate simplifies some calculations, the Ether concept was gradually abandoned and the foundation of Modern Physics based on the adopted postulates was established. In fact, after Einstein completed his theory of General Relativity in 1920 he reversed his opinion about the Ether, claiming "without Ether the General Theory of Relativity is unthinkable" [14]. The main reason was that he needed to explain logically the space curvature, one major result from General Relativity. While Einstein admired Maxell's theoretical ideas [13], he was not in favor of Maxwell's material Ether. Einstein's position is understandable because accepting the existence of a material Ether would have jeopardized his own theory of Special Relativity. Einstein's only argument against Maxwell's material ether was that "physicists failed to suggest a working model" [14]. At this time, supporters of Quantum Mechanics (predominantly based on mathematical logic) opposed Einstein's idea for any return to the Ether concept. For this reason even now Einstein's book "Sidelights of Relativity" is not mentioned in scholarly physics books and is even unknown to many university professors. In the 20th century, a number of contradicting problems accumulated. They appeared not only in Quantum Mechanics and the Theory of Special Relativity but also in other fields. They have been intentionally avoided and an idea was imposed that in the field of Quantum Mechanics and the Theory of Relativity, human logic fails. However, problems also plagued other fields like Particle Physics and even Cosmology. Today, accumulated astronomical observations put in doubt the Big Bang model of the Universe. As a result, an international Alternative Cosmology Group of scientists was officially established in 2004

[15] that started annual meetings called "Crisis in Cosmology Conference". Now with the advancement of new fields like nanotechnology, biophysics and cold fusion, new problems appeared that challenge the validity of established concepts of atomic models and elementary particles. The origin for all these problems is in the adopted space concept and related postulates. For the same reason, the UFO phenomenon is considered enigmatic and the official scientific community keeps a distance from this issue. Furthermore, accepting UFO as a real phenomenon will inevitably lead to shaking the foundation of Modern Physics at its fundamental level – the concept of space. In fact, this does not mean that the useful contributions of Modern Physics including Quantum Mechanics will suffer. The latter is a useful mathematical model, but its elements cannot be considered as real physical structures. It is known that Quantum Mechanics works only with a system of energy levels but in fact these energies are real signatures of interactions defined by real material structures of the elementary particles that build atomic nuclei. Such structures are not now envisioned by Quantum Mechanics because it relies only on mathematical models. The unveiling of these structures, including the structure of the space medium that defines the physical vacuum, opens a completely new window for advancement of our knowledge.

Now let us focus on one historical experiment, the results of which are wrongly used in scholarly physics books as a justification for abandoning the Ether concept. This is the Michelson-Morley experiment published in 1987 [16]. In fact, neither the conclusions of Michelson and Morley, nor the results from other later experiments disprove the existence of the Ether. For this reason it is useful to highlight some facts.

It is known that Michelson-Morley did not, in fact, receive a null result from their experiment, but something much smaller than expected. Assuming that the Ether frame is defined by the solar system (the concept of the Universe as a conglomerate of galaxies was still not known) they tried to measure the influence of the Earth orbital rotation (30 km/s) on the light velocity by using a sensitive Doppler shift interferometer with arms at 90 degrees. During the set of measurements that covered a minimum time of 24 hours, the interferometer was consecutively oriented in two orthogonal positions. There are a few reasons for not obtaining an expected result with this methodology: When using an interferometric method with a continuous light source, a number of effects could make the outcome of the experiment ambiguous. This must be taken into account not only for the Michelson-Morley experiments but also for other experiments provided later. Firstly, when assuming a motion in a medium, in this case the Ether, we should keep in mind that the Doppler shift occurs not only at the emitter but also at the receiver with an opposite sign. Secondly, the expected Doppler shift is contaminated by the relativistic clock rate change, discovered later by others [17] and by the Sagnac effect from the Earth rotation, also discovered later by the French physicist Georges Sagnac in 1913 [20]. Third, instead of the "Ether wind effect" from the Earth's diurnal rotation (0.458 km/s) and the orbital rotation (30 km/s), a larger effect from the Solar system motion around Milky Way (300 – 400 km/s) should affect the measurement. Fourth, a possible Fitzgerald contraction that might influence the result could not be identified. In their article Michelson and Morley mentioned that the same type of experiment would be repeated after 6 month, but at that time the experiment could not have a reference fringe position from the previous measurements. Everybody familiar with Michelson interferometers with a large path difference knows that it is extremely difficult to keep the stability of the fringe (for reference) even for 24 hours. In fact Michelson and Morley suggested a different experiment with a chopped light for measuring the effect of the Earth's orbital motion [16]. This experiment was not funded maybe because of the technological difficulties at that time. If it is eventually done, a surprising result might be observed because the solar system velocity is at least 10 times greater than the orbital Earth velocity of 30 km/s.

Here we must mention that not only the Michelson-Morley experiment but also some later experiments failed to detect our motion through the space medium (Ether) because the above-mentioned considerations were not taken into account. Some different types of experiments or observations, however, are able to detect a motion through the space medium. Amongst them are observations from the GPS system navigation. In this aspect, it is useful to cite the explanation of Roald R. Hatch, one of the pioneers of the GPS system. In the abstract of his article "Those scandalous clocks" [17] he writes:

Both VLBI (Very Long Baseline Interferometry) and GPS (Global Positioning System) indicate that earth-based clocks are biased as a function of their position in the direction of the earth's orbital velocity. The evidence for these biases is discussed, and the result is confirmed by comparison of earth-based clocks with millisecond pulsars. These clock biases are precisely such as to cause the speed of light to appear as "c" in the earth's inertial frame. This shows that the speed of light is not isotropic in the earth's frame and that the Lorenz transformation is only an apparent transformation that results from Selleri's inertial transformations combined with clock biases.

Now it is known that a number of cleverly designed experiments were able to detect the Earth's motion through space (see Appendix 1).

The Michelson-Morley experiment was done at a time when our understanding of the Universe as a huge conglomerate of galaxies was not yet established. At that time, the galaxies known as nebulas were regarded as star system formations inside of our Milky Way. Therefore, it is reasonable that a methodological misconception could have occurred. How could a motion through Ether be measured if the possible reference point were not yet envisioned? At that time, the so-called aberration in astronomical observations was a main issue of discussion, so it was considered that the stationary reference point of the Ether is located at the Earth or at the Sun. The solar system rotation around the Milky Way centre was not yet identified, so it was not regarded as a reference point for stationary Ether. Relativistic effects were also not well understood. In fact, the decision to abandon the Ether concept at that time was dictated not by experimental results but by the fast development of Quantum Mechanics and the emerging Theoretical Physics. The sad thing is that the conclusion about a possible methodological error in the outcome of Michelson-Morley experiment was realized 80 years later. By this time, a large amount of theoretical material based on the adopted space concept has already been developed and established as the status quo by the scientific authorities. Therefore, it is understandable why any proposed change or even correction of the adopted space concept and related postulates is now met with strong opposition. The established scientific community includes a large number of organizations financially supported by large government funds. So, they control not only the peer-reviewed journals but also have a strong influence on the media and policymaking. Since some fields of Theoretical Physics are now in stagnation, some challenges might be accepted, but not on the fundamental issue – the concept of space. This issue is intentionally avoided in such discussions. No conferences or ideas are supported that may eventually challenge the adopted space concept. Presently, the preservation of the status quo is not the policy of a single country but is an international one.

1.2. Brief summary of the BSM - Supergravitation Unified Theory (BSM-SG)

The treatise Basic Structures of Matter - Supergravitation Unified theory (BSM-SG) is based on an original idea about the structure of the physical vacuum. The suggested idea follows the recommendation of James Clerk Maxwell expressed in his famous theoretical work *A Treatise on Electricity and Magnetism* vol. II, section "A medium necessary":

In fact, whenever energy is transmitted from one body to another in time, there must be a medium or substance in which the energy exist after it leaves one body and before it reaches the other... Hence, all these theories lead to the conception of a medium in which the propagation takes place

The Basic Structures of Matter - Super Gravitation Unified Theory (BSM-SG) [1,4], unveils the relation between the forces in Nature by adopting the following framework:

- Empty Euclidian space without any physical properties and restrictions
- Two super dense fundamental particles able to vibrate and congregate
- A Fundamental law of Super Gravitation (SG) - an inverse cubic law valid in pure empty space.

An enormous abundance of these two particles, with energy beyond some critical level, is able to congregate into self-organized hierarchical levels of geometrical formations, based on the fundamental SG law. This leads deterministically to creation of space possessing quantum properties (known as a physical vacuum) and a galaxy as observable matter. All known laws of Physics are embedded in the underlying structure of the physical vacuum and the structure of the elementary particles. The fundamental SG law is behind the gravitational, electric and magnetic fields and governs all kinds of interactions between the elementary particles in the space of physical vacuum.

The underlying structure of the physical vacuum, called a Cosmic Lattice (CL), is of material origin. It is distinguished from the old Ether concept by a number of specific physical properties: high stiffness and pressure on more dense material objects at small scale; quantum mechanical and space-time features; and folding properties at the level of elementary CL node. As a result, the complex but well-defined behavior of CL structure permits explanation of all enigmatic phenomena in Particle Physics, Quantum Mechanics, Relativity and Cosmology. The new concept permits to explain the enigmatic space-time features and the relativistic effects by using a classical approach. Both, the special and the general relativistic effects a fully understandable when analyzing the behavior of the single elements of the CL space - the CL node and the motional behavior of an elementary particle, the electron for example. This analysis leads to definition of the basic physical parameters of CL space: a Static CL pressure, a Dynamic CL pressure, and a Partial CL pressure. The first one defines the Newtonian mass of the elementary particle (a mass equation is derived in BSM-SG, Chapter 3). The second one defines the Zero Point Energy (ZPE) related to the Electric and Magnetic fields. The third one is related to the inertial properties of the elementary particles. These features allow making analysis beyond Newton's laws of gravity and inertia and beyond the theory of Special and General Relativity.

Additionally, existence of two types of Zero Point Energy (ZPE) is found: static (ZPE-S) and dynamic (ZPE-D). The first is related to the Newtonian mass and the effects of General Relativity, while the second - to the Electric and Magnetic fields. One important feature of the CL nodes is their ability of self-synchronization with an identified signature - the Compton wavelength. This phenomenon is involved in the definition of the permeability, μ_0, and permittivity, ε_0, of the physical vacuum, which defines the constancy of the speed of light according to the equation $c = (\varepsilon_0 \mu_0)^{-0.5}$.

1.3. Considerations about matter, space, energy and gravitation according to the BSM-SG theory.

The Supergravitation (SG) law is defined by the expression (see §2.1. BSM-SG, Chapter 2)

$$F_{SG} = G_0 \frac{m_{01} m_{02}}{r^3} \tag{1.1}$$

where: F_{SG} - SG force; G_0 - intrinsic SG constant; m_{01}, m_{02} - SG intrinsic SG masses (different from the Newtonian mass); r – distance.

The proposed SG law is the most fundamental law in Nature. Its functionality is based on the ability of the fundamental particles and their formations to possess vibrational energy. The physics of the SG constant G_0 is discussed in Chapter 12 of BSM-SG, §12.A.6. We see that the SG low differs from the Newton's law of gravity by the inverse dependence of the forces from the distance and the value and sign of gravitational constant. This means that the SG forces are extremely strong at microscopic distances. The intrinsic SG constant also may change the sign, which is important for understanding the underlying material structure of the space. The participating SG masses are also much denser than the atomic masses. An example of the observed signatures of SG law is the Casimir forces detectable between two solid objects with polished surfaces at close distance. While the attractive Casimir forces has been detected for more than 30 years, a detection of repulsive Casimir-Lifshitz forces was recently reported [52].

Amongst the convincing experimental proofs for the existence of underlying structure of space as a medium defining the fundamental laws of physics is the detection of our absolute motion through space by laboratory experiments. This was proved by number of experiments, many of them published in peer reviewed scientific journals (see Appendix 1).

The Basic Structures of Matter - Super Gravitation Unified Theory permits:
(a) Understanding the fundamental relation between matter and energy
(b) Understanding the structure of the physical vacuum - Cosmic Lattice (CL) and its static and dynamic behavior permitting the definition of the space-time concept
(c) Understanding the physical relation between the Gravitational, Electric and Magnetic fields
(d) Solving the boundary condition problem for the photon as a quantum wave in a structured space of the physical vacuum
(e) Explaining the rules and effects of Quantum mechanics and General and Special Relativity by a classical approach (solving an old existing problem for the missing relation between them)
(f) Unveiling the physical structure of the electron, its oscillation properties and quantum features.
(g) Unveiling the physical structure of the elementary particles, the atomic nuclei and real quantum orbits. Understanding the cause of radioactivity.
(h) Unveiling the correct interpretation of the Einstein's formulae $E = mc^2$.
(i) Recognizing the existence of hidden space energy of a non-electromagnetic type as a primary source of nuclear energy
(j) Building an alternative Cosmology without contradictions based on the new space concept and reinterpretation of the observations

(k) Unveiling the levels of matter organization in the Universe based on geometrical formations in hierarchical orders.
(l) Deriving expressions showing the relation between the known physical constants and unveiled structural parameters of the physical vacuum and the elementary particles, including their mutual interactions
(m) The proposed model of two fundamental particles and one fundamental SG law provides an excellent opportunity for computer modeling of the unveiled geometrical formations and their rich vibrational properties.
(n) Unveiling the existence of longitudinal waves and their properties
(o) Showing that Newton's laws (about the universal gravitation and inertia) and Einstein's Special and General Relativity appear as special cases of the BSM - Super Gravitation Unified theory in a CL space environment.

1.4. Gravitational and inertial mass

According to the present status of the contemporary physics, the mass is associated with matter, so it is considered as unchangeable. Its value can be estimated by measuring the gravitational forces (gravitational mass) or by measuring the applied force for acceleration or deceleration (inertial mass). Both masses appear equal and their gravitational and inertial properties are given by the Newton's laws. We will refer this type of mass as a Newtonian mass.

According to BSM-SG, however, the Newtonian mass is not equivalent to matter. It is an attribute of the matter - a parameter that is used for estimation of matter quantity in a normal (non-disturbed) space environment, known as physical vacuum. The Newtonian mass is a valid parameter of the matter only if the object is immersed in CL space (physical vacuum), but this space is something completely different from a pure empty space. We simply could not reach or create a pure empty space. The Newtonian mass depends on the state condition of the CL space and if the normal CL space parameters are changed the mass will also change.

The BSM-SG unveils the material structure of the stable elementary particles, such as proton, neutron, electron and positron built by the same subelementary particles (twisted prisms) but arranged in helical structures with hierarchical order. The internal space volume of the helical structures contains internal lattice build by the same subelementary particles (twisted prisms) but much denser (about 1000 times) than the CL structure because it has a different spatial geometry. As a result, it is impenetrable by the CL structure, so the latter exercises a pressure on it. The analysis of the static and dynamic behavior of the elementary CL node under SG law allowed unveiling three types of pressure that the CL space exhibits in respect to material objects – from elementary particles to macro objects. This was possible by using the identified oscillating material structure of the electron and its quantum interaction with the CL space [2]. The analysis allowed derivation of the major CL space parameters and expressing them by the familiar physical constants (see BSM-SG, Chapter 2 §2.14 and Chapter 3):

- **Static CL pressure,** P_S - a pressure exercised on impenetrable volume of the elementary particle structure, defining the Newtonian mass of the elementary particle (see §3.13.3 Chapter 3 of BSM-SG).

$$P_S = m_e c^2 / V_e = h v_c / V_e = 1.3736 \times 10^{26} \quad [N/m^2] \quad (1.2)$$

where: P_S - is the static CL pressure, m_e – mass of electron, c – speed of light, h – Planck constant, V_e – impenetrable volume of the electron structure

Expressed only by the physical constants:

$$P_S = \frac{4 h v_C^4 (1-\alpha^2)}{\pi \alpha^2 c^3} \quad [\frac{N}{m^2}] \quad (1.2.a)$$

Using the Static CL pressure <u>the mass equation (1.3) for a stable elementary particles</u> is formulated:

$$m = \frac{4 h v_C^4 (1-\alpha^2)}{\pi \alpha^2 c^5} V_{ep} \quad [kg] \quad \textbf{mass equation} \quad (1.3)$$

where: V_{ep} – is an <u>impenetrable volume</u> of the elementary particle, h – is the Planck constant, v_c - is the Compton's frequency, c – is the speed of light, α - fine structure constant

- **Dynamic CL pressure,** P_D ~ (related to the Zero Point Energy of Dynamic type and responsible for the electrical and magnetic fields and the quantum behavior of the elementary particles):

$$P_D = h v_c / (c S_e) = 2025.8 \quad (N/(m^2 Hz)) \quad (1.4)$$

where: h – Planck constant, v_C - Compton frequency, $\tilde{S}e$ – impenetrable surface of the electron structure

- **Partial CL pressure,** P_P – related to the confined motion of the electron with one of its quantum velocities, in which the signature of the fundamental Fine Structure Constant α plays a role (see Table II in Appendix 2).

$$P_P = P_S \alpha v / c \quad \text{where } v \text{ is the particle velocity} \qquad (1.5)$$

For a moving electron, multiplying Eq. (1.5) by the electron's <u>impenetrable volume</u> V_e and having in mind the equality $hv_c = m_e c^2$ we obtain

$$\vec{E}_{IFM} = P_P V_e = hv_c \alpha \vec{v} / c = \alpha c m_e v \quad \text{[Nm]} \qquad (1.6)$$

The vector E_{IFM}, called an **Inertial Force Moment**, allows to estimate the deviation energy of the folded CL nodes, displaced from their normal positions at velocity v. It can be scaled for a moving proton (neutron) using the volume ratio between FOHSs of electron and proton, which is in fact equal to their mass ratio. Consequently it can be also scaled to a neutral atom, molecule and even to a solid body. The validity of Eq. (1.6) was verified with the Newton's Law $F = ma$ for the case of electron mass (§10.4.1.A, Chapter 10 of BSM-SG).

For a particle motion in which the velocity changes with the time the change of the Inertial Force Moment vector will be

$$\Delta E_{IFM} = \alpha c m a \quad \text{where: a - acceleration} \qquad (1.7)$$

Let us analyze a free fall of a particle, for example an electron. Then $a = g$ (gravitational acceleration), so: $\Delta E_{IFM} = \alpha c m g$ The gravitational potential at the initial moment is: $U_G = GMm/R$.

Dividing ΔE_{IFM} on U_G we obtain

$$U_G / \Delta E_{IFM} = R/(\alpha c) \quad \text{(s)} \qquad (1.8)$$

Eq. (1.8) is with a dimension of time, so we may write: $R/\alpha c = t$ and $R = \alpha c t$. Since all expressions are in system SI: $t = 1$ (s). Let us estimate R for a Compton's time: $t_c = 1/v_c$, where v_c is the Compton's frequency equal to 1.2355×10^{20} Hz.

$$R = \alpha c / v_c = 1.7706 \times 10^{-14} \, (m) \qquad (1.9)$$

The value of (1.9) appears equal to the helical step of the electron structure s_e, which is twice the small radius r_e of impenetrable volume of the electron structure (see Fig. 2.1 and Appendix 2 p. 144). This confirms the BSM-SG concept that the inertial property of a moving particle is defined by the deviation of the CL nodes around the impenetrable volume of the elementary particle. To do this they partially fold and deviate with rotational spin momentum that carries energy. The latter is responsible for the kinetic energy of the moving particle.

The mass equation (1.3) is valid for the stable elementary particles, such as electron, positron, proton and neutron, which are the building blocks of the matter. All they are built by material helical structure as identified by BSM-SG theory. The mass equation shows that the mass of the elementary particle could be changed if some of the parameters of the physical vacuum are changed. This conclusion is valid for atoms, molecules and solid body (see BSM-SG Chapter 10). In fact h, v_c and c in the mass equation (1.3) are commonly dependent, so the change of mass will affect more probably the speed of light or particularly the refractive index of the physical vacuum.

Conclusions:
(1) The **CL nodes fold and deviate** around the small electron radius of the moving electron - an indication of the **inertial interaction** of the electron with the CL space.
(2) The folded CL nodes spin. They take energy during the particle acceleration, contain it during uniform motion and return it during the particle deceleration. Therefore, the energy preservation is conserved.
(3) The derived result is transferable to any elementary particle, atom and molecule by normalizing their mass to the mass of the electron. Therefore, it must be valid also for a solid body.
(4) Because the gravitational and inertial masses are involved in derivation of Eq. (1.7) and (1.8), the derived result of Eq. (1.9) indicates that both, the gravitational and the inertial mass of a material body could be changed by proper modulation of the parameters of the CL space (physical vacuum). This is important conclusion for the properties of the field propulsion mechanism discussed later.

1.5. Longitudinal waves

The unveiled structure of the photon wavetrain (a neutral quantum wave) was described in BSM-SG, Chapter 2, §2.10.4 – 2.10.5) with a solution of the boundary conditions problem, which is important for preservation of the photon energy during its travel. It was found also that the momentum of Poynting vector is a result of wavetrain helicity. In a normally generated EM wave, the E and H vectors are not only orthogonal but having also a small longitudinal component in the direction of propagation, which is in accordance with the existed mathematical treatment. However, in some special cases of generation of EM wave, this longitudinal component can be increased significantly.

The Longitudinal waves envisioned by Lord Kelvin [39] are firstly observed by Nikola Tesla 100 years ago. From the point of view of Classical Electrodynamics, the existence of LWs is apparent only if using the original forms of Maxwell's equations; i.e., the quaternion form. This is now theoretically proved by a number of theoreticians [21] (Van Vlaenderen K. J. and Waser A. (2001)); K. P. Butusov [22]. LWs have quite different properties than the ordinary EM waves. Presently experiments demonstrating the emission and receiving of LWs are reported in peer-reviewed journals [40].

The existence of CL space structure means that longitudinal waves (LWs) are possible as compression-like waves, different from the familiar EM waves.

Understanding the wavetrain structure of the photon (and EM waves) permits us to guess what could be the configuration of the longitudinal waves (LWs). They should possess a strong longitudinal component. One way to understand these waves is to imagine that they contain counter rotating E and H vectors. While EM waves, for example, could be generated by a solenoid, one might wonder what might be the configuration of a solenoid or antenna for generating LWs.

Additional technical considerations exist for generation and reception of LWs, which are discussed elsewhere in this book. LWs are naturally generated by a lightning. The LWs have a large penetrating capability and converts to broadband EM waves, known as transients. These transients are able to pass through conventional EM filters and could destroy sensitive equipment. BSM-SG theory envisions three types of LWs depending of the way they are generated and the conditions of their propagation:
(a) Isotropic LWs
(b) LWs embedded in EM waves
(c) LWs in closed magnetic lines

Isotropic LWs may propagate with a superluminal velocity but they attenuate fast with the distance.

The LWs embedded in EM waves may appear hidden for an ordinary EM receiver. However, they are able to carry independent information and are propagated with the light velocity.

The LWs in closed magnetic lines may propagate with superluminal velocity.
The three types of LWs are particularly useful for the special applications discussed later in this chapter.

1.6. On possibility of changing the mass of a solid object according to BSM-SG theory

The possibility of controlling the gravitational and inertial mass of a solid object is not envisioned by the contemporary Modern Physics, so this issue has not been discussed in the mainstream journals and media. The attempt to access this issue, while relying on the space concept adopted 100 years ago, usually leads to speculative ideas accompanied by highly abstract mathematics without useful practical recommendations. Advance in this field couldn't be achieved unless the problem is accessed from a new concept of the physical vacuum.

When the mass is estimated by its weight we may consider its possible change as decreased gravitational attraction from the Earth. In a more general case, however, we may consider the change of the inertial mass, which is estimated by the acceleration (or deceleration). In this case one may assume for analysis that the object is very far from any gravitational field.

BSM-SG theory is able to provide an understandable relation between gravity and inertia (Chapter 10 of BSM-SG) and also between the gravitational, electric and magnetic fields, using the derived static and dynamical properties of CL space. This allows to predict a possible change of the gravitational and inertial mass of the object if manipulating some of the parameters of the CL space (physical vacuum). The physical understanding of the phenomenon permits also envisioning the technical methods for such manipulation of the mass.

Now let us see what defines the inertial mass according to the BSM-SG theory. All stable elementary particles such as electron, positron, proton, and neutron are comprised of First Order Helical Structures (FOHS) containing internal lattice much denser that the CL space structure (Chapter 2, 3 and 6 of BSM-SG). When the elementary particle moves through CL space, its impenetrable FOHSs are wrapped by the displaced and partly folded CL nodes, since the CL node geometry is flexible (see Chapter 10 of BSM-SG). This physical phenomenon defines the inertia of the stable elementary particle. If the velocity approaches the speed of light, the elementary particle experience increasing resistance. <u>The reason for</u>

this is that the rate of separated and partly folded CL nodes approaches the CL node resonance frequency. This effect is at the basis of the relativistic increase of the inertial mass according to Einstein's Special Relativity. Investigating the relativistic motion of the electron the relativistic gamma factor was derived in §3.11.A1, Chapter 3 of BSM-SG. The described phenomenon is valid also for atoms, molecules, gases and solids.

In order to understand the inertial properties of an elementary particle or a solid object in CL space we must have a reference frame. From the analysis of the astronomical observations in Chapter 10 and 12 of BSM-SG, it becomes evident that the space of the Milky Way (and other galaxies) could be considered as an absolute reference frame. This is confirmed by a large number of appropriately arranged experiments measuring the vector of our motion in absolute space (See Appendix 1). Some of these experiments are made in a laboratory and they usually detect the larger component of our absolute motion, which comes from the solar system motion. Since this velocity is in the range of about 300 - 400 km/s, it is completely inconsistent with the Big Bang model of expanding Universe, according to which the Universe expands with a velocity approaching the speed of light. This is one of many problems of the Big Bang model, which became inconsistent with the results from the accumulated cosmological observations. From the other hand the detection of our motion in absolute space is in excellent agreement with the scenario of an Alternative Cosmology concept presented in Chapter 12 of BSM-SG, as a consequence from the alternative concept of the physical vacuum. It not only demonstrates that the galactic redshift is not of Doppler type but also offers an explanation, which agrees with many observed cosmological phenomena. It is wrongly written in some physics textbooks that Edwin Hubble, the discoverer of the galactic redshift, is discoverer of "expanding Universe". The truth is that he did not consider that the observed red shift is of Doppler origin, so he never accepted the concept of expanding Universe.

Understanding the existence of an absolute frame of reference is an important issue for further analysis of our motion through space.

In the analysis of the dynamical behavior of the CL nodes (Chapter 2 of BSM-SG), they are regarded as Phase Locked Loop (PLL) oscillators. It is well known that such oscillators, possessing a proper (intrinsic) frequency, are easily synchronized by phase. BSM-SG analysis envisioned the existence of ZPE waves as groups of CL nodes synchronized by the phase of the SMP vector propagating with the speed of light. The average length of the ZPE waves is equal or multiple of Compton's wavelength $\lambda_C = 2.426 \times 10^{-12}$ (m). This is the distance that the phase of the SPM vector propagates with the speed of light per one SPM cycle of the CL node. The period of the SPM cycle is equal to the Compton's time $t_c = 0.809 \times 10^{-20}$ (s). The ZPE waves appear as continuously recombining, so they are responsible for the equalization of the ZPE-D and for the space-time properties of the physical vacuum. They are also involved in the definition of the permittivity and permeability in vacuum (free space), which are responsible for the constancy of the velocity of light.

How does the Newtonian gravitation of a heavy astronomical body like the Earth attract a material object? The SG forces between the Earth and the object are propagated through the CL space structure. More specifically, the Super Gravitational field is propagated along the *abcd* axes of the CL nodes (Chapter 2 of BSM-SG), which are always aligned and separated by automatically supported small gaps (the latter phenomenon is defined by the specific properties of the SG field, which are discussed in Chapter 12 of BSM-SG). At the same time, every CL node vibrates with its proper (intrinsic) resonance frequency

$$v_R = 1.0926 \times 10^{29} \ (Hz) \quad (1.10)$$

The SG field of the prism is characterized by the propagation of a SGSPM vector, the frequency of which is obtained by division of the primary Planck frequency, f_{PL}.

$$f_{PL} = (2\pi c^5 / Gh)^{1/2} = 1.855 \times 10^{43} \ (Hz) \quad (1.11)$$

where: c – light velocity, G – gravitational constant, h – Planck constant

The mechanism of frequency division is embedded in the intrinsic matter structure of the prisms forming the CL node. It is based on stable frequency modes defined by the intrinsic mechanical properties of stable geometrical structures from which the twisted prisms, forming the CL nodes, are built (see §12.A.4 and §12.A.6., Chapter 12 of BSM-SG). According to this concept the frequency of the SGSPM vector (propagating inside the prism) is higher than the CL node resonance frequency v_R (1.10). Consequently, the CL node resonance frequency provides an attenuation effect for the long-range propagation of the SG field in CL space. This conclusion from BSM-SG theory is in agreement with the theoretical derivation by H. E. Puthoff in his article "Gravity as a zero-point-fluctuation force" [24]. Starting from Planck's frequency and using a hypothesis suggested by the Nobel Prize Russian academician A. Saharov, Puthoff derived the law of Newtonian gravitation by attenuating the higher frequencies. The conclusion that the frequency of SGSPM vector propagated within the prism is higher than the CL node resonance frequency is confirmed also by the pendulum experiment of Dobromislov [25], when analyzed by the BSM-SG theory. In the experiment of Dobromislov, a torsion pendulum placed in a vacuum chamber exhibits asymmetrical displacement of the middle position in dynamical mode (clock-wise and counter clock-wise motion) in respect to static position. According to BSM-SG this effect is caused by the twisting asymmetry between the (external) material structures of the protons and neutrons from one side and the electrons from the other. This asymmetry causes a small difference in the attenuation of the SGSPM vector (the carrier of the SG field in CL space), so this difference is detected in Dobromislov's experiment.

The above considerations lead to the following conclusions:

(A). The long-range propagation of the inverse cubic SG field in the CL space appears as a Newtonian gravitational field, which is inversely dependent on the square of the distance)

(B). When analyzing the SG propagation through CL space, the oscillating CL nodes could be regarded as static due to their intrinsically small inertial factor in pure empty space (see Chapter 2 and 12 of BSM-SG)

(C). The resonance frequency of the CL node imposes some attenuation effect on the propagation of the SG field through the CL space

Feature (B) is very important for understanding the properties of the inertial frame formulated in Special Relativity. It becomes apparent that a formulation of inertial frame makes sense only if it is attached to some object possessing a mass and for the points of space where the local gravitational field is larger than the external one. The discovery of features A, B and C permitted a successful analysis for unveiling the structure of the magnetic lines and the wavetrain shape of the photon (§2.10.3 and §2.10.4 Chapter 2 of BSM-SG).

1.7. Change of gravity by disturbance of the CL space surrounding the object.

From features (A) and (C) at the end of previous section it is evident that the propagation of the SG field between two material objects (Newton gravity) will be facilitated if the CL space between them contains selfsyncronized microdomains. In other words, the propagated SG forces will be stronger in the case of synchronized CL nodes, in comparison with the case of non-synchronized nodes. In a normal space environment the uniformity of the ZPE-D energy is kept by the ZPE waves, which are spontaneous result of self-synchronized CL space microdomains. From BSM-SG analysis in Chapter 2 it is evident that the selfsinchronization of the CL nodes creates synchronized microdomains with the length equal to whole numbers of Compton's wavelengths $\lambda_C = 2.426 \times 10^{-12}$ (m). The two main parameters of the CL structures, the CL node distance and the proper frequency v_R are kept constant namely by the self-synchronization (ZPE waves). This defines the very constant appearance of the light velocity (one CL node distance for one cycle of v_R). Obviously the self-synchronization keeps constant the two known fundamental parameters of the physical vacuum: permittivity, ε_0, and permeability, μ_0. That's why the velocity of light is defined by the equation $c = (\varepsilon_0 \mu_0)^{-0.5}$.

Now let us consider the self-synchronization for two cases:
(1) In a CL space domain far from any gravitational field
(2) In a CL space domain in a gravitational field.

In case (1) the distance between the CL nodes will be kept only by the ZPE of dynamic type. Its measurable parameter is the temperature of 2.72K estimated by the Cosmic Microwave background. This parameter is estimated only for a large distance. It cannot be estimated for the space surrounding the Earth.

In case (2) a second factor – the gravitation of the massive object will slightly affect the both parameters – CL node distance and the CL proper frequency. This causes a small shrink of the local CL space with a small change of the CL node proper frequency v_R. Consequently, the Compton's frequency (time) playing a role of a time etalon in Quantum Mechanics will be slightly changed. These space-time physical effect is namely behind the space curvature predicted by Einstein theory of General Relativity. Einstein intuitively felt that the Ether is needed for explanation of the space curvature, but he did not find a working physical model of the material Ether. In his book Sidelight of Relativity, he said: *Without Ether, the General Relativity is unthinkable.*

Consequently, Newtonian gravitation regarded as a propagation of the SG field will depend on the permanent existence of the ZPE waves. **What will happen if the synchronization of the CL nodes is disturbed?** Obviously, the propagated strength of the SG field will decrease, which means a decrease of Newtonian gravitation between Earth and the object. Therefore, for manipulating the gravitational forces some kind of disturbance of the CL node synchronization is necessary. Such manipulation could be from some natural event or intentionally invoked.

When considering the disturbance of CL self-synchronization we must take into account one important parameter - a restoration time after removing the disturbance. Obviously, it should be defined for some unit volume. This means that body with different size will have a different restoration time after a gravity disturbance.

Additionally we must consider one important property of the CL space: the self-adaptation of the CL structure to gravitational disturbance and its ability to restore the self-synchronization to the previous normal level after a strong gravitational pulse. The physics of these CL property is based on the intrinsic features of the interaction of the CL nodes: since they operate in a pure empty space they don't have an inertial properties like the material structures of the elementary particles which operates in CL space (physical vacuum). From this considerations the following conclusion could be made:

Conclusion: The disturbance of the self-synchronization is more effective in the case of a gravitational pulse.

The above conclusion is supported by a number of experiments. Some of them are natural, others are artificially invoked. One example of a natural disturbance of self-synchronization is the effect of a local gravitational disturbance measured during a solar eclipse by a Chinese and German group of scientists and published in Physics Review D [23]. The measured data are shown in Fig. 1.1.

Field Propulsion by Control of Gravity

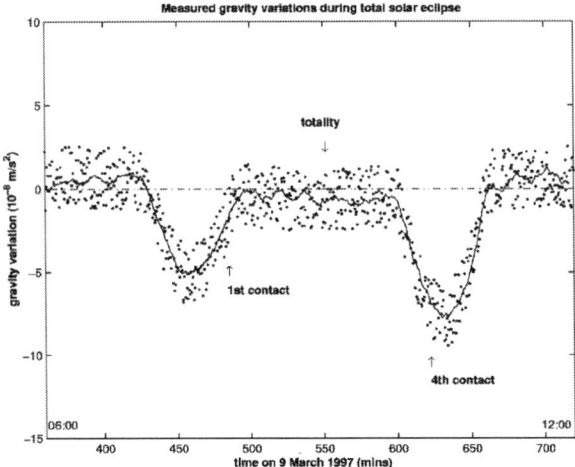

Fig. 1.1. Variation of vertical gravity measured during the solar eclipse on 9 March 1997. Courtesy of Quian-shen Wang et al, [23]

The interpretation of the published data is given here according to the BSM-SG concept. It is clearly visible that the large gravity variation occurs not during the time of total eclipse but when one edge of the moon disk (in a line of sight) crosses the first edge of the solar disk and when the other edge of the moon disks crosses the second edge of the solar disk. Furthermore, in both cases the local gravity decreases. This confuses completely the interpretation based on the officially adopted space concept, according to which if there is some "gravitational shield" it must be stronger at the time of total eclipse and the possible deviations are expected to be with opposite signs. However, the measured result is in complete agreement with the above mentioned considerations for disturbed CL space synchronization: the disturbance leads to diminishing the local gravity and it occurs during the moment of the larger gravitational change considering the line of sight alignment of the Sun, Moon and Earth.

Gravity wave detectors have been suggested by different researchers. One simple but effective gravity detector has been suggested by Gregory Hodowanec [26]. Fig. 1.2 shows its electrical circuit diagram. Hodowanec found that capacitors with some type of dielectric are sensitive to gravitational disturbances.

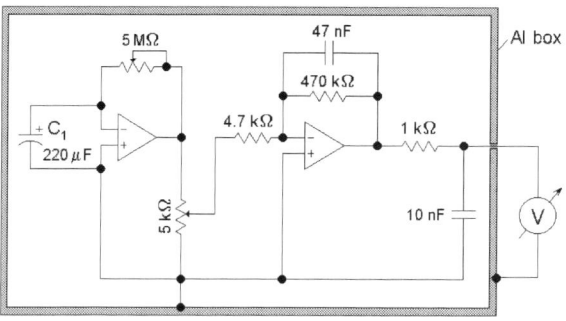

Fig. 1.2. Electrical circuit diagram of the Hodowanec gravity waves detector [26]. The sensitive element is the capacitor C_1.

The whole circuit shown in Fig. 2.2 is enclosed in an aluminum box. The sensitive element in fact is the dielectric of the capacitor C1 and Hodowanec selected it to be of a proper type. He tested the detector with rotation masses enclosed in shielded boxes of different metals placed in a close proximity to his detector. Hodowanec provides his own theory called "Rhysmonic Cosmology" suggesting an explanation of gravity by using vectors attached to the masses but going in every direction [27]. Then the field defined as "density of Rhysmonic vectors" appears to change inversely proportional to the square of distance. In fact, the effect of CL node synchronization from the body mass is similar, while a density of a similar field is explainable by the degree of self-synchronization affected by the massive body. In this sense, the BSM-SG concept provides an understandable explanation of the physical principle involved in the Hodowanec gravity detector. The change of gravitational forces by gravity waves disturbs temporally the CL node synchronization that affects the permitivity of free space ε_0. This causes a small change in the dielectric constant of the capacitor C_1. As a result, the capacitance suddenly changes. Since it is charged by the bias voltage of the operational amplifier IC_{1A} a pulse is detected. Furthermore, the sensitivity of such a detector is large because the charge on the capacitor C_1 is proportional to the amplified voltage, so during the gravity pulse some kind of positive feedback occurs. In such a way, this simple circuit appears to be a sensitive detector of gravitational disturbances.

Using his detector, Hodowanec was able to make unique observations of astronomical events. Amongst his observations is a measurement of the disturbances from the Sun line of sight position passing over local noon meridian. The methodology of the measurement and the observed data are shown in Fig. 1.3.

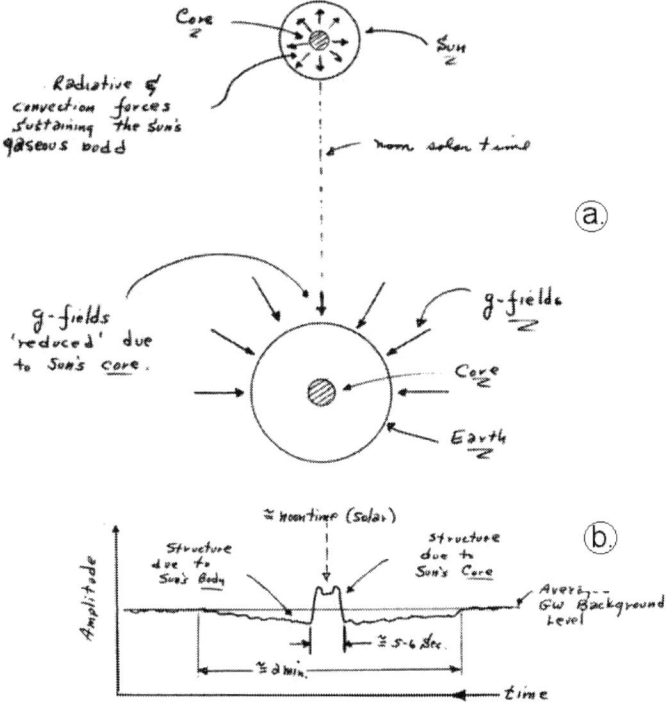

Fig. 1. 3. Measuring of the noon pass of the Sun (over the local meridian). Courtesy of Gregory Hodowanec [33] (a) a sketch showing the possible existence of highly dense cores inside the Sun and the Earth; (b) Noon-time pass measurements on 12-22-1986.

Firstly, we must pay attention to the central core zone noted by Hodowanec as a core. The existence of such a core with a non-spherical shape and density much greater than the density of any heavy metal is predicted by the BSM-SG theory, Chapter 10 §10.14 (Magnetic field hypothesis for astronomical object) and Chapter 12§12B.6.4. (Analysis of pulsar observations). According to BSM-SG, every star or planet with a mass larger (or denser) than the planet Mercury may have such a super dense core. Observations of shock waves from Earthquakes really indicate the existence of such a core with physical properties quite different from the rest of the Earth's body [28]. The observations of seismic waves show also some directional differences in their propagation in respect to the geographical and magnetic polar axes. This is in complete agreement with the BSM-SG prediction that the Earth contains a super dense central structure that is responsible for the Earth magnetic field.

According to BSM-SG, Gregory Hodowanec was the first who was able to measure the gravitational signature of the earth central core. He also suggested what could be its approximate size based on a number of observations. The central core signature is apparent from the plot of Fig. 1.3. In contrast from the gravitational disturbance by the total solar eclipse (see Fig. 1.1), Fig. 1.3.b. shows the disturbance with a larger temporal resolution. Firstly, the Hodowanec experiment indicates that the central cores have a stronger gravitational disturbance than the other parts of the Sun and Earth. Secondly, the gravitational disturbance from the central core is different from other part of the body. This disturbance also exhibits a finite time constant.

Hodowanec also observed a local gravity disturbance during solar eclipses. Another interesting result he found from such observations is that the measured effect precedes the line of sight alignment by about 8 minutes. This is just the time for light from the Sun to arrive at the Earth. <u>Consequently, the gravity waves propagate instantaneously, m</u>uch faster than the speed of light. This confirms Newton's statement that gravitation is instantaneous. This is contrary to the Einstein's statement that the gravitation propagates with the speed of light. Hodowanec was also able to detect gravitational waves coming from the center of our Milky Way galaxy. Now it is known that there is a "black hole" with enormous mass here. According the BSM-SG this is a superdense primordial matter, while the matter of the observable galaxy is from the same origin but much more rarefied.

One famous Russian scientist, Nikolai Kozyrev, also confirmed by astronomical observations that gravitational waves propagate instantaneously. With a detector not sensitive to EM radiation and put in a focal point of a reflective telescope (with an aluminum type of mirror but optically blinded input aperture) he was able to detect signals from stars many light years away in positions where they should be seen optically after many years. After some routine astronomical calculations, it became evident that the detected signal arrives instantaneously, vastly exceeding the speed of light.

Kozyrev also provided laboratory experiments demonstrating a temporary decrease in the gravitational mass of a solid body. He stressed solid bodies of different materials by shaking or vibrating, then put them on a sensitive balance to

Field Propulsion by Control of Gravity

measure their mass. He selected vibrations with a large harmonic at some selected frequency that depended on the material. After such treatment, he immediately put the body on a sensitive balance and measured the curve of restoration to their normal weight. The effect appeared only for bodies made of non-elastic materials. Figures 1.4 and 1.5 show the restoration of the body mass after it has been submitted to proper vibrations.

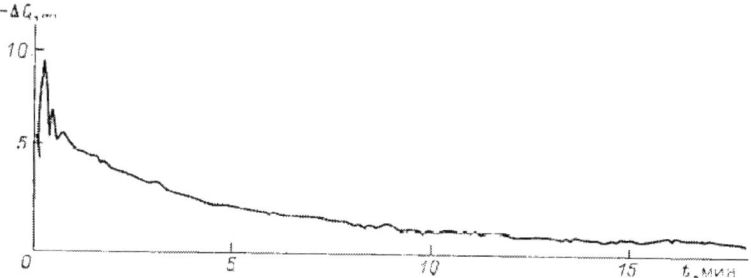

Fig. 1.4. Weight restoration of a metal box with weight of 108 (g) after multiple shocks from a steel ball inside. Measurement by a technical weight meter. Courtesy of N. Kozirev [29]

The time resolution of the restoration curve in Fig. 1.5 indicates that the restoration process is concurrent over some average value. This is in agreement with the concept of restoration to normal CL space synchronization.

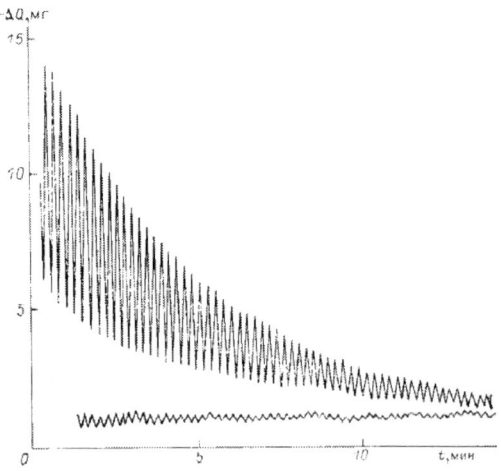

Fig. 1.5. Weight restoration of a heavily hammered copper plate using analytical weight meter. Courtesy of N. Kozirev [29]

28

1.8. Field Propulsion by unidirectional change of the Newtonian mass of a body

Now let us consider how the synchronization of the CL nodes, or in other words the normal state of ZPE waves, could be disturbed in the space surrounding the solid body in order to decrease its gravitational (and probably inertial) mass. Without entering into details, we may envision the following methods:
- disturbing the CL node synchronization by emission of longitudinal (scalar) waves (LWs)
- disturbing the CL node synchronization by gamma rays
- disturbing the CL node synchronization by producing a conflict of magnetic line directions, based on the dynamic properties of the CL nodes involved in magnetic lines (MQ SPM is discussed in Chapter 2 of BSM-SG).
- disturbing the electric field of accumulated charge by producing a conflict based on the dynamic properties of the CL nodes involved in the electric lines (EQ SPM is discussed in Chapter 2 of BSM-SG).
- <u>disturbing the CL node synchronization by using the oscillating properties of the electrons</u>

Considering the case of using LWs, we must keep in mind that they contain a longitudinal component resulting from a compressible effect of CL space in which the strong SG forces are directly involved. On the one hand, the LWs interact directly with the strong hidden ZPE-S energy, so they may carry much more energy than the ordinary EM waves and for this reason they are very penetrative. On the other hand, they may effectively disturb the CL node synchronization for a finite time interval, during which the following effects will occur:

(1) decreased gravitation between the object and Earth
(2) disturbed EM waves in the space surrounding the object
(3) a blurred appearance of the object when imaged from a distance

Effect (1) is the necessary one for manipulating the gravitational mass. At the same time, the disturbed synchronization affects the permittivity and permeability of the surrounding space, so the EM field and the propagation of light in this zone will be also disturbed. This causes the side effects (2) and (3).

In Chapter 10 of BSM-SG, it was shown that the inertia of a solid object is related to the <u>integral inertial momentum of displaced and folded CL nodes, which is</u> expressed by the <u>force moment</u>

vector, E_{IFM}, given by Eq. (1.6) shown in §1.4. This vector depends on the number and velocity of the displaced folded CL nodes, which defines the inertial properties of a particle, atom, molecule and solid object. Formulated in such a way this vector is able to describe both - the involved energy and the inertial properties of any kind of motion (uniform, rotational or accelerated).

The E_{IFM} vector will get a directional moment if the speed of light is affected by some kind of asymmetrical disturbance of the CL node self-synchronization around a particle with mass m.

In uniform linear motion (no acceleration and changing the direction) the Force Moment vector is constant. The momentum of folded CL node at the entrance is transferred to the unfolded CL nodes at the exit, so no external force is needed. In a uniform rotational motion, only the direction of the force moment vector is changing, so a centrifugal acceleration is felt. In the case of linear motion with acceleration, the magnitude of the force moment vector is continuously changing, so a continuous force is felt. In a case of accelerating motion in a curve trajectory, both the magnitude and the direction of the force moment vector are changing.

Now let us consider an asymmetrical disturbance of the CL space self-synchronization around a neutral atom or molecule. In a normal non-disturbed CL space the Newtonian mass is given by the mass equation (1.3), which is defined as a Static CL space pressure on the impenetrable volume of the particles involved in the atom or molecule. If the Static CL pressure is affected asymmetrically, it is evident that its mass will be affected by getting also a directional moment. This effect must be valid also for a body formed of atoms or molecules.

The validity of the mass Equation (1.3) and the inertial property Equations (1.6) and (1.7) propagate to atoms, molecules and to a solid object. The latter is regarded as integral entity of stable elementary particles.

Conclusion: <u>Asymmetrical disturbance of the CL node self-synchronization around an elementary particle, a neutral atom, a molecule or a solid object will cause a change of its gravito-inertial mass according to Eq. (1.3) and a unidirectional non-inertial displacement according to Eq. (1.6) and (1.7).</u> From both equations it is evident that the common parameter c – speed of light should be also affected if achieving an interaction with the Compton frequency $\nu_C = 1.236 \times 10^{20}$ Hz, which is one of the basic parameters of the physical vacuum.

Normally all CL nodes, which are in the pathway of the translational motion of the particle, partly fold and wrap the impenetrable FOHSs of the particle, the volume of which defines its Newtonian mass. This CL node behavior is valid for the cases of moving atoms, molecules and solid objects. Now, we must emphasize two important features of the (partly) folded CL nodes. First, they carry the inertial motion energy of the moving particle; second, they do not have strong connections between themselves like the normal nodes connected into CL structure. Then we may conclude:

(D). A fraction of folded nodes could be deviated and guided by a strong magnetic field with an appropriate configuration

(E). The deviated and guided folded nodes will cause a displacement of the object without experiencing a force as in normal acceleration (or at least - a reduced force). We may call this **a Field Propulsion effect, which can be regarded as caused by a propulsion Force field of new kind** (it was referred as manipulative displacement in BSM-SG, Chapter 13).

(F). If the maximum velocity of a solid object is within the range of our solar system velocity around the Milky Way centre (including the Earth orbital motion), we are sure that the equivalence between the gravitational and inertial mass will be preserved.

Feature (F) means that both the inertial and the gravitational mass will appear equal but smaller in the case of Field Propulsion. This leads to the following important conclusions:

(J). A spacecraft using a Field Propulsion Mechanism will be able to make a sharp turn or reversal of its direction without feeling an excessive acceleration.

(H). The acceleration of the object in the case Field Propulsion will require less force and energy.

1.9. Hypothesis of Field Propulsion mechanism

The predicted propulsion mechanism differs physically and technologically from the mechanism of jet propulsion systems. The new physical aspect of the predicted propulsion is based on change of the Newtonian gravitational (inertial) mass.

Below is a briefly presented hypothetical version of a spacecraft using the predicted propulsion mechanism, while focusing only on the physical principle and some secondary effects. From the considerations discussed in the previous section, it is apparent that the geometrical shape of the spacecraft is important.

Figure 1.6 shows one example of the overall shape and the main functional blocks of a spacecraft using the predicted propulsion mechanism. However, this is not the only option of such type of spacecraft. (In Chapter 4, a more detailed technical approach is described).

The spacecraft enclosure 1, made of appropriate material, must transmit the LWs from the radiators 3. At the same time, it may have magnetic properties (made of proper material layers) in order to sustain a superstrong magnetic field 2 surrounding the spacecraft.

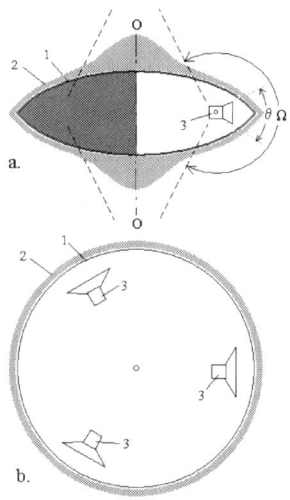

Fig. 1.6. Spacecraft with a Field Propulsion mechanism based on manipulation of the gravitational (inertial) mass. Views a. and b.
1 - spacecraft envelope, 2 - super-strong magnetic field,
3 - radiators of longitudinal waves, θ - instant angle of radiation, Ω - coverage angle (by rotation of the radiator)

After the CL space disturbance a finite time interval is need for its restoration to normal condition. This is theoretically envisioned as a restoration of the CL node self-synchronization or relaxation time constant of CL space, the existence of which is apparent also from the analysis of some experiments and observed phenomena. The finite value of the restoration time permits the radiators 3 to operate in a burst mode sequence. The duration and the repetition burst rate in this mode must correspond to the spacecraft's

Chapter 1

maneuvering characteristics. The radiators may also rotate to cover the field angle, or alternatively a larger field angle could be obtained if the spacecraft enclosure posses an appropriate refractive feature. The radiators should also have a phase and intensity control of the emitted LWs. The super-strong magnetic field also could be guided near the surface of the spacecraft enclosure by combination of magnetic and paramagnetic materials. Since the magnetic field and LW may have some conflicting features, the researchers and designers of such spacecraft must be acquainted with the BSM-SG theory.

A spacecraft with a properly designed shape, material and propulsion system will be capable of achieving a fast displacement (acceleration), while the crew inside will not feel the acceleration.

The trip to a planet or a distant solar system will contain three phases: an acceleration phase, a constant velocity phase, and a deceleration phase. The maximum velocity and acceleration may have an upper limit defined by the intrinsic features of the CL space (physical vacuum).

Let us consider the most conservative option, relying on our motion through the CL space of the Milky Way with a velocity that is in the order of 300 km/s. (It is apparent from Chapter 12 of BSM-SG why this velocity range is taken for reference). Since the Earth's orbital velocity is about 30 km/s, we are sure that we cannot have some unknown biological effect at a velocity less than 30 km/s, relative to Earth. Obtaining this velocity with an acceleration of 9.81 m/s^2 will take about 41 min, so it is insignificant for interplanetary flights. At one of the closest positions to the Earth (for example in November 2005), the distance to Mars is about 70 x 10^6 km, so that a one-way trip there should take about 27 days.

In this conservative scenario, we excluded the possibility of much faster acceleration and higher velocities because of the uncertainty about unknown biological effects. Such a restriction may not exist, but this could be verified only by future experiments. If so, from the point of view of BSM theory, we may consider only one restriction, which is related to the electron velocities in quantum orbits. If we take into account that normal human temperature is below 40 C, then the electrons will be in lower level orbits. We may take the 3rd subharmonic orbit as an upper limit for which the electron energy is 1.51 eV, corresponding to a velocity of 729.7 (km/s). (Note: BSM atomic models show that the orbital electron velocities in the heavier elements are not much different from those of hydrogen, because the higher energy levels known from the atomic spectra come not from the kinetic energy of the electron but from the SG field potential energy). Then we must consider two cases:

(a) the folded CL node velocity inside the spaceship is equal to the spaceship velocity (the crew feels the acceleration)

(b) the folded CL node velocity inside the spaceship is managed to be smaller within acceptable limits and not dependant on the spaceship velocity (the crew does not feel large acceleration)

In case (a), even for an average distance to Mars of about 238×10^6 km, the acceleration (or deceleration) phase with 9.81 m/s^2 is less than a day, while the phase with a constant velocity of 729.7 (km/s) is less than 4 days.

In case (b), keeping the folded node velocity within some acceptable limit depends on the spaceship design. This option is suitable for distant space travels. While the biological species might be vulnerable, robotic spacecraft may operate in conditions of higher acceleration and velocity in order to shorten the time of space travel.

Now let us discus some of the features of the spacecraft illustrated in Fig. 1.6. The zone around the axis OO is a zone of interference of the emitted LWs, while the round zone between the large circular section and the axis OO is suitable for sensors monitoring the motion. The Field Propulsion efficiency (achieving a larger acceleration) will be larger in a direction with a small cosine value between the velocity vector and the plane of the larger sectional area of the spacecraft because the folded CL nodes must be deviated at a smaller angle. When moving horizontally, the spacecraft will be able to move in a zigzag fashion with large sudden changes of direction and acceleration. For fast acceleration with a vertical angular component in respect to Earth, the spacecraft must be initially properly tilted and then accelerated. It is evident that a spacecraft with such a shape is suitable for space travels between planets with large acceleration and velocity. For interstellar travels, a different shape of the spacecraft will be more appropriate. This will be discussed in Chapter 4.

1.10. Physical effects accompanying the Field Propulsion mechanism

When moving in the Earth's atmosphere, the described spacecraft will appear blurred, especially when accelerating. The air molecules or the gas surrounding the spacecraft will be ionized and will emit broadband radiation. If the surrounding CL space is uniformly disturbed, a gradual lens effect will take place in which the whole or part of the spacecraft may look blurry or semitransparent. In that case, some radars may not be able to detect the spacecraft. Radar pulses, in the way they are generated, contain some LWs embedded in the EM pulse. Radar pulses with a stronger LW component will more reliably detect such a spacecraft.

The disturbed synchronization of the CL nodes destroys temporarily the magnetic protodomains, which in fact are embedded in the magnetic lines. Consequently, the weak magnetic field of the Earth will be affected locally. The affected space region might be significantly extended beyond the spacecraft body and it will contain fragmented domains. The disturbed self-synchronization requires a finite time for restoration depending on disturbed volume and degree of disturbance. So, when the spacecraft is passed on, the local Earth magnetic lines of this zone will be restored within a finite time interval. They will have various paths trying to avoid the disturbed domains. The detectable effect will be such that a magnetic compass will be unstable with erratic behavior. When the spacecraft is very close to some electronic instruments, their operation might be temporarily disturbed due to the disturbed EM properties of the surrounding zone. Staying outside in a proximity to operating spacecraft (during landing or standby) should be avoided. The LWs may affect the energy storage mechanism of the biomolecules as discussed in Chapter 11 of BSM-SG.

Other non-conventional micro-effects on material structure and their properties may take place near the spacecraft.

The spacecraft using a Field Propulsion effect does not need any atmosphere or to be near a massive astronomical object. Its propulsion mechanism might be even more effective in deep space; however, it will need a rarefied gas envelope around it – a feature discussed in next chapters.

1.11. Signature of some characteristic features of the new propulsion mechanism in the observed UFO cases.

A large number of UFO related publications, containing description of the observed physical features have been analyzed from the BSM-SG point of view. While this issue has not been officially in the focus of mainstream science, the author was careful in the selection of the published material. Among the reliable sources are the proceedings of the Workshop "Physical Evidence Related to UFO Reports" held in Pocantico Conference Center, Tarrytown, N.Y., Sep 29 - Oct 4, 1997; and the book *The UFO*

Enigma, a new review of the physical evidence, by Peter A. Sturrock [31], a distinguished astrophysicist and Emeritus Professor. Another useful book is "Unconventional Flying Objects a scientific analysis" by Paul Hill [30] a UFO researcher (1909-1990) who worked in NASA.

The main and side physical effects predicted by the BSM-SG theory concerning Field Propulsion are in excellent agreement with the observed physical phenomena discussed in these books and other reliable sources.

Peter Sturrock describes one well-documented observation made by a government airplane mapping the coast in Costa Rica on September 4, 1971. In one of the consecutive video frames, an object appears as shown in Fig. 1.7.

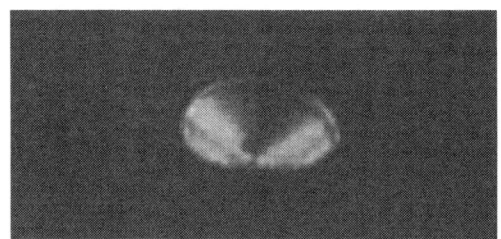

Fig. 1.7. Picture of observed object (Adapted from P. A. Sturrock, The UFO enigma, a new review of the physical evidence, (p. 202, Fig. 25-17) [31]

The object is pictured from above against a uniform background of water. Instrumental records of orientation, coordinates, and local time accompany the picture. The following text is extracted from the Peter Sturrock's book [31]:

First, the disk image appears to possess light/dark shading that is typical of a three-dimensional object that is illuminated by sunlight.
Second, the generally triangular dark region on the right-hand side of the disk cannot be a solar shadow cast by the (assumed) opaque disk from the right-hand side. If the disk is an opaque, flat conical section of revolution (the dark spot being the tip of the cone) and if the right side is tipped upward, then the entire surface of the disk should be dark. It is more likely that the light and dark regions are surface markings...
Forth, while the right-hand edge of the disk image is in very sharp focus, the left-hand edge is diffuse and appears to be an irregular boundary which almost transits the light of the background in a transparent manner. It is of interest to note that the general orientation of this left-hand boundary of the image runs north and

south rather than being parallel with the visible longitudinal axis of the disk.
Fifth, the entire image is in sharp focus suggesting that (a) the shutter speed was fast, (b) the disk was not moving relative to the Earth background, or both. It is known that the exposure lasted 1/500 seconds, which would "stop" a slowly moving object but not necessary a fast-moving one.
... The 4.2 mm length of the image is equivalent to an object 210 m in length, or 683 feet.

This observation is only one among a large database of documented materials gathered during the past 50 years, but the described effects are very common. Presently a large database of UFO observations exists. Despite lack of government support, a number of organizations with members of emeritus professors and distinguished engineers organize annual scientific meetings with workshops about UFO phenomena or related topics. Amongst them are the Society of Scientific Exploration [32], Integrity Research Institute [33], Center for UFO Study [34] and many other institutions over the world. Credible long-time UFO researchers with scientific background have gathered large collections of UFO observation, many of them accompanied with recorded physical data. Amongst them are Paul Hill [30], Peter Sturrock, [31], Richard Haines [59], Robert Dean [60] and many others. One of the long time researchers devoted to UFO and endorsing the idea of its reality is the nuclear physicist Stanton Friedman [35]. Presently many individual researchers collect and investigate new pictures or videoclips.

1.12. Field Propulsion side effects

The Field Propulsion could be accompanied by some unusual side effects that are not present in the classical jet propulsion system. According to the BSM-SG, a real UFO spacecraft leaves an unique optical signature in the surrounding space that can be captured on a good quality picture. The researcher William Treurniet, who is well acquainted with the BSM-SG theory, developed an image processing technique for identification of one particular signature that permits to identify a real UFO with some degree of confidence [36]. His research shows that spacecraft in the earth atmosphere are accompanied by a weak optical effect of torus-like formations in the surrounding air space. For conventional aircraft the effect could be a result of turbulence, so it is very weak. In the UFO case the effect appears more apparent because the mass of air molecules surrounding the object is also changed. However, the flight of UFO is not accompanied by turbulence as in the conventional aircraft, as investigated by Paul Hill [30] and the affected zone may exist longer. For this reason, the torus-like formations in UFO cases might be more apparent.

From the point of view of the currently adopted space concept, the observed physical affects accompanying UFO seem quite mysterious. As a result, many speculative "explanations" have been suggested, such as: "materialization" and "dematerialization", other dimensions, wormholes, human hallucination, etc. These pseudo-explanations are completely wrong. Reliable physical records exist indicating that the UFOs are real objects. What is missing so far is the physical explanation of the observed phenomena.

The BSM-SG predictions for changing the gravitational mass of an object are also in agreement with the observed and documented but so far unexplained phenomena referenced as Hutchison effects. These effects were discovered accidentally by John Hutchison from BC, Canada, while trying some unconventional experiments. In fact, John Hutchison was able to demonstrate two different physical effects: The first is a temporary weight loss of a material object and the second is unusual change of the structure and shape of a solid body (metal or dielectric) [37,38,62]. The weight loss has been witnessed by a number of individuals and credible researchers. Videoclips of moving or floating objects demonstrated by John Hutchison are available by Internet (videoclips by John Hutchison). Some of his experiments are described in a recent book published by G. Hathaway [62]. John Hutchison obtained these effects many times mostly in a closed room environments where a number of different apparatuses were properly set up. Amongst them were a Naval RF generator, HV equipments like Van de Graaff generator, HV transformers and Tesla Coils of different configuration. [38]. From the point of view of BSM-SG, the weight loss is a result of conflict between high intensity DC and AC fields in a tiny area and perhaps due to some standing waves. For this reason the effect is difficult to reproduce in other places. The effect could be possibly invoked by a Heterodyne Resonance Mechanism described in Chapter 2. The second effect - a structural change - appears as bending or crushing of objects made from metal or other material, or penetration of one material into another without heating or melting (for example, a metal object into a wooden object at room temperature). Evidently, the object that underwent such treatment obtained a changed structure. John Hutchison possesses many samples with a changed structure. From the point of view of BSM-SG, the structural change is possible because of changing the bonds between molecules and atoms in a solid body. In fact, it is an undesirable side effect, but it is very important that it be taken into account by future designers of a spacecraft using the described propulsion mechanism.

The author of this book accidentally obtained one case of a weak Hutchison effect of structural change during his experiments. It is shown in the picture of Fig. 1.8.

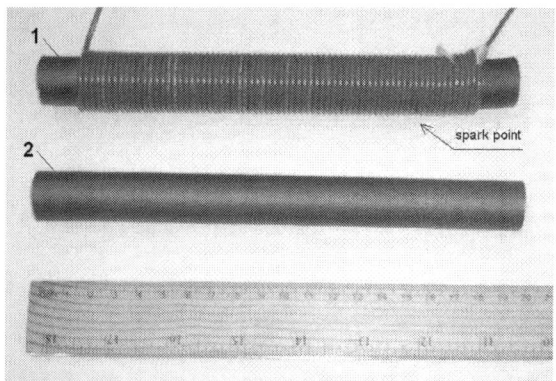

Fig. 1.8. The ferrite rod 1 wound with wire is bent and deformed without any cracks. The bare rod 2 is identical and shown for comparison.

The ferrite rod 1 wound with wire was accidentally bent and deformed without any cracks during one of the experiments provided by the author of this book. The bare rod 2 is identical and shown for comparison. The wound rod 1 was used as a solenoid in the HV circuit for activation of the Heterodyne Resonance Mechanism (discussed in the next chapter). During the experiment, a spark to ground occurred accidentally and the rod distorted without the presence of any heat. The dissipated energy from the spark was no larger than 30 watts per 3 seconds. The rod is with OD 20 mm, ID 12 mm and length of 20 cm. The bending point is 45 mm from one end. The ID also got an elliptical shape with a diameter ratio of 0.92.

The accidentally obtained Hutchison effect shows that unwanted destruction of the material may occur if the Field Propulsion technique is not correctly applied. In fact, the system for Field Propulsion must be considered as an inseparable part of the spacecraft design.

The fact that UFO spacecraft may leave a temporarily unsafe zone has been investigated by the Canadian researcher Wilbert Smith. He says:

"...*the field surrounds the saucers in order to hold them up. In order to produce the gravity differentials, time field differentials are necessary to operate the ships. These sometimes produced field combinations which reduced the strength of materials to the point where they were no longer strong enough to carry the loads that the materials were expected to carry. Now as we know, aircraft — particularly the military type aircraft — are built with a rather small factor of safety, and if they fly into a region of reduced binding, the material is no longer strong enough to carry the load, and the craft simply comes apart*" [61].

Wilbert Smith suggested a device that is able to detect space zones of reduced "binding force". The device was tested in a few airplanes and showed some positive results. However, the lack of understanding of the physical principle was the reason the device was not implemented. The working principle of this device is completely understandable from the point of view of BSM-SG theory. Wilbert Smith also mentioned that large unsafe space zones were caused by nuclear tests. They could appear as long-time "drifting holes". According to BSM-SG theory these are large zones with severe disturbances of CL node self-synchronization.

CHAPTER 2. Heterodyne Resonance as a physical mechanism for invoking Field Propulsion

The disturbance of the CL nodes self-synchronization in the volume surrounding the object (spacecraft) will cause a unidirectional change of its gravito-inertial mass as a new kind of force field as discussed in section §1.6 of this book. In this chapter we describe the physical aspect of one approach for affecting the self-synchronization of CL space (physical vacuum).

2.1. Accessing the Compton frequency by using the oscillation properties of the electron

The new space concept permits unveiling the underlying material structure of space (physical vacuum) with its quantum properties and Zero Point Energy (ZPE) on one side, and the material structure of the elementary particles on the other. It also shows the possible quantum interactions between them and explains them in a classical way.

From the BSM analysis (discussed in the article "Brief Introduction to BSM theory and derived atomic models" [6] and presented in detail in Chapters 2, 3, 4, 5 of BSM), we see that the ZPE-D (dynamic type of Zero Point energy) is related to the CL node dynamics, which is characterized by the following two frequencies:

CL node proper resonance frequency:
$$\nu_R = 1.0926 \times 10^{29} \quad \text{(Hz)} \tag{2.1}$$

SPM vector frequency = Compton frequency:
$$\nu_C = 1.2356 \times 10^{20} \quad \text{(Hz)} \tag{2.2}$$

Frequency ν_R is directly related to the propagation of photons (or any kind of EM radiation) with the velocity of light, whereas the frequency ν_C is directly connected to the permeability of free space, which is one of the physical parameters responsible for the constancy of the speed of light (see §2.10.3 and § 2.10.4 from Chapter 2 of BSM-SG).

From the material presented in §1.6 and §1.7 of this book we see that to affect gravitational mass we must provide some kind of self-synchronization disturbance in the surrounding CL space. To do this, we must access one of the super-high frequencies ν_R or ν_C.

While v_R is practically unreachable, the frequency v_C could be accessed by using the quantum properties of the electron. In fact this frequency is directly related to the self-synchronization between the CL nodes.

The unveiled material structure of the electron (Chapter 3 of BSM-SG) exhibits all of the electron's known physical properties: Compton frequency and wavelength, anomalous magnetic moment, Quantum Mechanical spin, and gyromagnetic moment. Furthermore, additional important features are unveiled permitting explanation of the quantum mechanical properties and interactions using a classical approach. The BSM model of the electron is published in Physics Essays – a peer reviewed journal devoted to fundamental questions in Physics [2] (see Appendix II). Figure 2.1 shows the electron structure with its dimensions and oscillation properties. (The internal lattices of this structure are not shown here for simplicity).

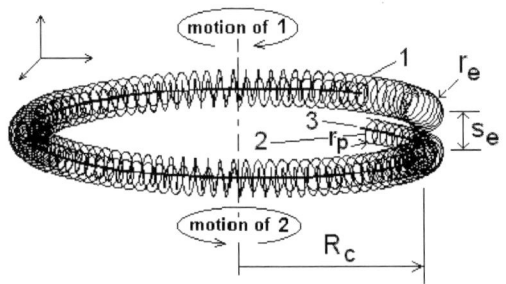

Fig. 2.1. Oscillating electron (the internal lattices are not shown)
1 – electron external helical structure, 2 – internal helical structure (positron), 3 – central core, $R_C = 3.8616 \times 10^{-13}$ (m) - Compton radius

The external helical structure 1 is of prisms having a twisting that modulates the CL nodes of prisms with the same handedness, and it appears as a negative charge. The internal helical structure 2 is of prisms of opposite handedness providing a positive charge. The core 1 is also from prisms with twisting corresponding to a negative charge. In the curved cylindrical space inside the structure 1 and 2 there are denser lattices of type different than the CL space lattice (not shown in this figure). The internal lattice of the negative helical structure 1 allows the internal positive structure 2 to oscillate in conditions of ideal bearing. The identified proper frequency of this oscillation, called a first proper frequency is equal to the well-known Compton frequency. The second proper frequency is the oscillation of the structure 3 inside of the structure 2 and it is found to be higher (See Chapter 3 of BSM-SG) but its affect is weak. The SPM frequency of the CL node in the Earth's gravitational field is exactly equal to the first proper frequency of the electron (Compton frequency). The analysis of the suggested physical model of the electron in Chapter 3 of BSM-SG unveiled also one important feature: its helical step s_e is defined by the well known fine structure constant according to the expression

$$s_e = \frac{\alpha c}{v_c \sqrt{1-\alpha^2}} = 1.77 \times 10^{-14} \quad (m) \quad (2.3)$$

By definition: $R_C = 3.8616 \times 10^{-13}$ (m)

It is also found that:

$r_e = s_e g_e = 8.84 \times 10^{-15}$ (m) – small electron radius

$r_p = 5.89 \times 10^{-15}$ (m) – small positron radius

where: $\alpha = 7.2973525$ - fine structure constant, $g_e = 2.0$ - is the gyromagnetic factor of the electron

The material structure of the electron is the simplest one compared to other stable elementary particles. However, it possesses rich quantum interaction properties with the CL space (physical vacuum) due to its geometry and oscillation properties (for more details see Chapter 2 and 3 of BSM-SG).

BSM-SG analysis of the confined motion of the electron in CL space unveiled the following important features:

Feature (A): The first proper frequency of the oscillating electron is equal to the SPM frequency of the CL node, which in the Earth gravitational field is the well-known Compton frequency, v_C.

Feature (B): When the electron moves with a velocity corresponding to energy of 13.6 eV, its quantum interactions with the CL space (physical vacuum) is optimal. <u>This is so because the phase of the oscillating electron matches the phase of the CL node oscillations described by the SPM vector propagated with a light velocity.</u> Other suboptimal (quantum) velocities are also possible but the above-mentioned phase match occurs not for every rotational cycle of the electron. They correspond respectively to energies of $13.6/n$ (eV), where n – is a whole number. For suboptimal confined velocities, the phase match between the oscillating electron and the phase of the propagated SPM vector occurs every n cycles. Therefore the quantum interaction strength diminishes with the number n.

A brief explanation of features (A) and (B) is provided below.

Figure 2.2 shows the motion of the electron for two cases:
Case a. Motion with Optimal and Suboptimal velocity
Case b. Motion with Super-optimal velocity. In both cases the electron trajectory is helical but it interact differently with the CL structure (physical vacuum) due to its oscillation properties.

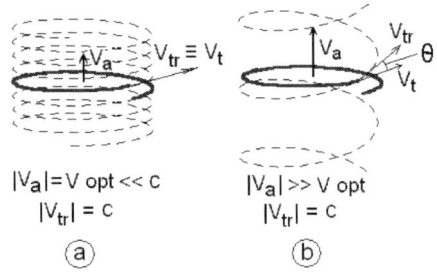

Fig. 2.2 Electron confined motion with an optimal velocity (Case a.) and with a velocity higher than the optimal one (Case b.)

V_A – axial velocity

V_t – tangential velocity component on a helical trace

V_{tr} – tangential velocity of rotating electron

V_{ALT} - alternative motion velocity of the oscillating internal helical structure of electron (see Fig. 2.1)

c – light velocity

As shown in Fig 2.2, even for velocity larger than the optimal one (13.6 eV), no point on the moving and rotating electron structure can exceed the velocity of light (because the displaced CL nodes could not fold faster than the frequency v_R, which together with the node distance defines the light velocity). For this reason, the trajectories of both cases are different. We must not forget, however, that the internal helical structure of the electron (see Fig 2.1) oscillates with a Compton frequency. Consequently, we must consider that the vector of tangential velocity contains two components: V_{tr} - a constant component and V_{ALT} – an alternative component (superimposed on V_{tr}) defining the oscillation with a Compton frequency caused by the alternating reversal motion of the internal helical structure (structure 2 in Fig. 2.1) of the electron.

While the magnitude of V_{ALT} is much smaller than the magnitude of V_{tr}, it is responsible for the quantum interaction of the electron with the CL nodes, which means a quantum interaction with the physical vacuum.

Case a. The step of the helical trajectory is exactly equal to the helical step of the electron structure s_e, shown in Figure 2.1. Therefore, this type of motion is completely screw-like. However, we may distinguish two subcases of case **a.**

Subcase a.1 The tangential velocity is equal to the velocity of light:

$V_t = c$

Subcase a.2. The tangential velocity is less than the velocity of light

$V_t < c$

In case **a.1** the axial velocity, V_A, is equal to the product of light velocity, c, and the fine structure constant $\alpha = 7.29735 \times 10^{-3}$

$V_A = \alpha c = 2.1877 \times 10^6$ (m/s)

This corresponds to electron energy of 13.6 eV. In other words, the fine structure constant, one of the most fundamental parameters (this is quite evident also from the BSM-SG theory), is embedded in the dynamics of the electron structure as a ratio between its axial and tangential velocities for its screw-like optimal confined motion corresponding to energy of 13.6 eV). At this velocity the internal structure of the electron oscillating at the Compton frequency meets the CL nodes oscillating at the same Compton frequency (described by the SPM vector propagating with light velocity), always with the same phase.

The phase match between the first proper frequency of the electron (Compton's frequency) and the SPM frequency (Compton's frequency in Earth gravitational field) appears at every cycle of electron oscillations for velocity $V_a = \alpha c$, corresponding to electron energy of 13.6 eV.

<u>Consequently, at velocity corresponding to 13.6 eV the oscillating electron and the propagating SPM vector are both phase locked. In this case the quantum interaction of the oscillating electron with the space of physical vacuum is optimal. This velocity is further referred as optimal quantum velocity. It is equal to the</u>

electron velocity in Bohr atomic model for the principal quantum number $n = 1$.

In Case **a.2** the step of the electron trajectory is still equal to the electron structure parameter s_e. However, its velocity could be n times smaller than the optimal velocity $V_a = \alpha c$. The phase match between the oscillating electron and the propagated SPM vector occurs on every n cycles. Then we have

$$V(n) = \frac{\alpha c}{n}, \text{ where n – whole number} \qquad (2.4)$$

The classical equation for energy in (eV) is
$$E = 0.5 V^2 / q \quad \text{where q – charge of electron} \qquad (2.5)$$
Substituting V in (2.5) with (2.4) we get
$$E = 0.5 [V(n)]^2 / q \qquad (2.6)$$

Investigating the length of the electron orbits for hydrogen contributing for line spectra (Chapter 7 of BSM-SG) it was shown that when *n* increases V decreases while the electron trajectory length is still the same. This corresponds to the motion case of the electron as shown in Fig. 2.2.a. However, the phase match between the electron proper frequency and the CL node frequency will not occur for every cycle of the electron proper frequency but *n* times rarer. For this reason this type of screw-like motion of the electron is called a motion with a subharmonic number, n, which in fact is equal to the principal quantum number. The relation between this number *n*, the axial velocity of the electron, its energy in (eV) and the Quantum efficiency is given in Table 2.1. The Quantum Efficiency 1 corresponds to the optimal confined motion of the electron in which its quantum interaction with the CL space (physical vacuum) is optimal.

Table 2.1

n	Axial velocity (m/s)	Energy (eV)	Quantum efficiency
1	2.187×10^6	13.6	1
2	1.094×10^6	3.4	1/2
3	7.292×10^5	1.51	1/3
4	5.469×10^5	0.85	1/4
5	4.375×10^5	0.544	1/5

Chapter 2

It is well known that the energy level of 13.6 eV corresponds to a ground state in the hydrogen atom, and if the atom is cool enough, the electron may stay at this level a longer time. BSM analysis unveiled the physical reason for this - the optimal quantum interaction with the physical vacuum (see the analysis in §3.11.A, Chapter 3 of BSM-SG). Therefore, the electron energy of 13.6 eV is not a feature defined by the hydrogen atom but is an intrinsic dynamic characteristic of the electron moving in CL space. In the Bohr atomic model of hydrogen, the energy 13.6 eV appears from the model, but this model in fact is mathematical – not a real physical one. The problems of the Bohr model were widely discussed at the time when Quantum Mechanics was developed.

Case b. The step of the helical trajectory is larger than the electron step s_e and the oscillating vector V_{ALT} does not coincide with the tangential vector V_t. For this reason, the quantum interaction with the CL space (physical vacuum) in this case is weaker.

The quantum efficiency of the electron moving with a velocity larger than the optimal one diminishes with the velocity. Using the suggested electron model, it is found in BSM-SG chapter 2 § that the quantum efficiency in this case is given by the equation

$$\eta = (1 - V^2/c^2)^{1/2} = 1/\gamma \qquad (2.7)$$

where: V – is the electron velocity, γ - is the relativistic factor

A number of other physical parameters of the electron excellently fit to the physical properties of the suggested electron model (Chapter 3 of BSM-SG). We must not forget that the circumference length of the electron structure is equal to the Compton wavelength $\lambda_C = 0.024263 \times 10^{-10}$ (m), but the same parameter is also a characteristic parameter of the CL space that possesses quantum properties. As a space parameter, the Compton wavelength, λ_c, is equal to the path passed by the phase of the SPM vector, propagated at light velocity for one SPM cycle (a Compton time, t_c): $\lambda_C = ct_c$. The BSM analysis unveiled that the <u>Compton wavelength defines the smallest possible length of the magnetic line, so it is a characteristic parameter of the Zero Point Waves.</u> These waves exist permanently in the CL space (physical vacuum) as self-synchronization between individual CL nodes. They are responsible for the definition of the permeability of the free space, μ_0. They are also behind the equalization of the ZPE-D energy through the space (related to the background temperature of the CL space 2.72K – a measurable parameter from the Cosmic Background Radiation). In Case **a.**, the whole electron structure makes a complete rotation for one SPM cycle (Compton period). This condition is not satisfied for Case **b.** and the difference becomes larger when V_t is larger than the optimal velocity corresponding to electron energy of 13.6 eV.

<u>Conclusion (C): The SPM (Compton) frequency of the CL space is theoretically accessible if using the oscillating properties of the electron defining its quantum interaction with CL space domains. The quantum interactions have maximums at velocities corresponding to the energy set given by the expression $13.6/n^2$ (eV), where n – is integer. The optimal interaction is at 13.6 eV (n = 1), in which case the oscillating electron (with Compton frequency) and the oscillating CL node frequency (SPM = Compton frequency) are phase locked. The strength of the quantum interactions decreases sharply when *n* increases.</u>

The Conclusion (C) is an extremely important result from the BSM-SG analysis. In fact by accessing the superhigh Compton frequency using the electrons we may affect the ZPE waves or, in other words, we may disturb the self-synchronization of the CL nodes in some particular space volume.

Now let us see how such a disturbance could be realized. Using the oscillating features of the electron (its rotational motion and simultaneous oscillations with its proper frequency v_C) we may achieve disturbance of the ZPE waves by a frequency type of interaction with the CL node frequency v_C. Firstly, it is apparent that we must have a strong quantum interaction between the moving electron and the CL space, which is satisfied for electrons with energies of 13.6 eV (optimal, or first harmonic motion) and lower (3.41, 1.51 and so on – sub-optimal, or subharmonic motion). In the case of 13.6 eV (first harmonic motion), the oscillating electron is phase locked to the phase of the SPM vector (propagating at light velocity) which is involved in the ZPE waves. In subharmonic motions at 3.41 eV, or 1.51 eV and so on, the phase lock is of lower order and the quantum interactions with the CL space are weaker. Obviously, we must assure a phase locking condition between the moving electrons and the CL nodes, and then invoke interactions

Chapter 2

that may work against this condition. This could be achieved by first assuring motion of electrons with optimal or suboptimal quantum velocity and then invoking a change in the direction of motion. In other words, we must use an alternating electrical field but its frequency could be much smaller that the Compton frequency.

2.2. Quantum mechanical spin of the electron

The oscillation properties of the electron's material structure permit explaining the QM spin of the electron in a classical way. It is related to the first proper frequency of the electron, which appears to be equal to the experimentally determined Compton frequency $\nu_C = 1.2356 \times 10^{20}$ (Hz). At the same time, the SPM frequency of the CL node is also equal to the Compton frequency (in Earth's gravitational field). When the electron moves with its optimal quantum velocity (energy of 13.6 eV), the phase of its oscillations matches the phase of the SPM vector propagated with the velocity of light. The phase match may occur either at the front end (with respect to the screw-like motion direction) or the back end (see Fig. 2.1). Since the internal structure 2 in fact is a positron, the magnetic moment in one case will reinforce the magnetic moment created by the electron, while in the other case will reduce it. Additionally, there is a possibility for a change in the upper coil handedness of the electron structure. In Fig. 2.1, the second order helix is shown with a left-hand handedness. However, the stiffness of the structure is not very strong and there could be a change to a right-handed helicity as a result of external magnetic conditions. In the second case, only the second order helical handedness is changed while the first order helical structure preserves its handedness. This means that the magnetic moment of the electron will be changed when the second order handedness is changing. The helical step will be the same because the two positive ends of the internal positron will repel each other if the helical step is less than the value of s_e given by equation (2.3). This is in fact the QM spin corresponding to +/- h/2, where h – is the Planck constant. In Chapter 3, an experiment is described demonstrating the spin flipping of the electrons when forming pairs with positive ions.

The material structure, oscillation properties and quantum interactions of the electron with CL space explains all known Quantum Mechanical properties of the electron including its QM spin, magnetic radius, anomalous magnetic moment, gyromagnetic factor and how its motion defines quantum mechanical orbits in atoms (BSM-SG, Chapter 7) and molecules (BSM-SG, Chapter 9).

2.3. Heterodyne Resonance mechanism and SARG effect

From the previous section we see that we can access the super-high Compton frequency of CL nodes $v_c = 1.236 \times 10^{20}$ Hz by invoking a proper oscillation motion of electrons using an electrical field with much lower frequency. This mechanism, discovered by the author of BSM-SG, is called a Heterodyne Resonance Mechanism. It is predicted and briefly discussed in Chapter 13 of BSM-SG. In this section we focus on the physics of this mechanism.

The physics of the Heterodyne Resonance Mechanism is illustrated by Fig. 2.3. where 1 – is the trajectory of the single ionized atom, 2 – is the helical trajectory of the bound electron, 3 is the magnetic field line of the electron moving in a helical trajectory, 4a and 4b are electrodes on which AC high voltage electrical field is applied.

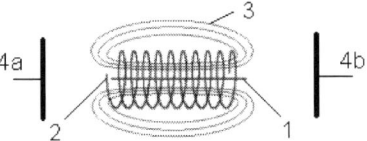

Fig. 2.3. Trace and magnetic field of electron bound to a positive ion forming an ion-electron pair. 1- positive ion trace, 2 –electron trace, 3 magnetic field from the electron, 4a and 4b – electrodes providing AC electrical field

Considering a moving positive ion with a trajectory 1, the bound electron will make a helical trace 2. If the positive ion motion is reversible, the bound electron will also make a reversible helical motion. Since the helical step of the electron's structure mentioned above is much smaller than the electron's Compton radius, the confined motion velocity of the electron moving on the helix 2 will be much greater than the ion velocity. This allows the electron to move in a helical trajectory centered on the bound positive ion with one of the electron's quantum velocities corresponding to energies of 13. 6 eV or 13.6/n eV (n – integer), while the velocity of the ion can be much smaller. It is well known that the magnetic moment of the electron is 658 times greater than the magnetic moment of the proton and 981 times greater than the magnetic moment of the neutron. At the same time, even the smaller mass element (the

Chapter 2

positive one-proton hydrogen ion), is 1836 time heavier than the electron, so it could be accelerated to a much lower velocity. Then the magnetic field of the bound system of single ion-electron will be predominated by the magnetic field created only by the electron. The magnetic field from an electron moving in a helix is much stronger than if it were moving in a straight line. Additionally, the magnetic fields of the neighboring ion-electron pairs interact constructively. As a result, a large number of ion-electron pairs move in a cluster.

The described mechanism is achievable for an actuator containing two electrodes separated by a proper gap and operated in a gaseous atmosphere with conditions for ionization. The ionization and the described alternating motion can be achieved by using a strong alternating electrical field or short pulses in the accessible RF range. In fact the applied electrical field must trigger the initial acceleration in the reversible motion of ion-electron pair. The bound ion-electron pair or more appropriately, cluster, will have its own oscillation proper frequency. The frequency of applied AC HV must be lower in order give a freedom of the process. The bound electron of the ion-electron pair moves in a helix with one of its most probable quantum velocities corresponding to energies of 13. 6 eV or 13.6/n eV. The proper frequency of the ion-electron pair oscillations depends on the type of working gas and the pressure. It may be adjustable in some limited range since the step of the electron's helical trajectory could be accommodated in some range. In some technical methods for invoking a Heterodyne Mechanism, the applied AC field could be synchronized with the proper frequency of the ion-electron pairs that will permit observation of the spin-flipping signature of the electron. The spin flipping may not occur for each ion cycle but for only a few cycles of ion-electron frequency. This is demonstrated by experiments discussed in the next chapter.

From the point of view of the unveiled electron structure, the spin flipping is a change of the phase of the oscillating electron structure (at the Compton frequency, see Fig. 2.1) by 180 deg with respect to the Compton frequency of the CL nodes. <u>At this particular moment an interaction momentum takes place between the electron and the CL space (physical vacuum).</u>

The practical realization of the Heterodyne Resonance mechanism is the creation of neutral plasma around a properly designed actuator. The physical process is characterized by the following consecutive phases:

- ionization of neutral atoms or molecules
- ions get acceleration
- formation of ion-electron pairs, each one formed by a single-charge positively ionized atom and a free electron
- ion-electron pairs get reversible oscillation motion triggered by strong electrical field of AC or pulse DC type and self-sustained by the magnetic field created by the bound electrons
- the frequency of reversible oscillation motion of ion-electron pairs is generally not the same as the frequency of the strong electrical field
- <u>the reversible motion of the electrons bound to positive ions is accompanied by spin flipping, however, it may not occur on every cycle but on every few cycles.</u>
- the interaction between the reversible moving electron and the physical vacuum appears at the moment of spin flipping

For optimization of the Heterodyne Mechanism the following considerations must be taken into account:
(a) The reversible motion of the ion-electron pairs can be disturbed by collision, so the process needs reactivation
(b) The process is more effective for electron-ion pairs in which the ion is from atom with lower atomic mass. In this case the magnetic moment of the electron highly predominates the ion magnetic moment.
(c) The frequency and strength of the applied electrical field that causes ionization and formation of oscillating ion-electron pairs must be properly selected. It depends on the type of working gas, the pressure and the distance between electrodes.
(d) An external magnetic field properly oriented in respect to the magnetic field of bound electrons might increase the efficiency.

The Heterodyne Mechanism takes place in EM activated neutral plasma excited by AC high voltage or by RF or microwave activation. A signature of positive ion-electron pairs is found in the analysis of a large number of experiments involving EM activated neutral plasma. It is apparent in experiments using different gases and at different pressure ranges – from a vacuum to a normal atmospheric pressure and even overpressure. In many published experiments, the researchers point out that the average electron velocity is in the range 3-10 eV, which is much lower than expected but they don't offer a satisfactory explanation for this. This range, however, is in good agreement with the predictions of the

Heterodyne Mechanism, since the estimated average electron velocity in fact is an average value of the electron velocities involved in the ion-electron pairs (mostly energies of 13.6 eV and 3.41 eV). The formation of ion-electron pairs must be done frequently due to unwanted collisions of ion-electron pairs with neutral molecules and other ions. In that respect, it is convenient to apply an AC HV field, which can be in the kHz frequency range (depending on pressure).

Practically, the Heterodyne Mechanism can operate in a wide range of pressure from a few mbar to atmospheric and even above atmospheric pressure. The formation of ion-electron pair clusters helps to increase the effective free path. Nevertheless, the effect strength is reduced in normal atmospheric pressure due to losses from collisions and the presence of negative oxygen ions. The efficiency is strongly dependent on the average free path between collisions. The collisions contribute to a significant fraction of the broadband high frequency spectrum, which is unwanted EMI noise. Only single ionized positive ions can participate in the working ion-electron pairs. Some negative ions significantly disturb the process.

The observed physical phenomenon of obtaining Field Propulsion is called Stimulated Anomalous Reaction to Gravity (SARG) effect.

The selection of working gas is also important, as this is evident from the above-mentioned considerations. Massines et al. [41] investigated an atmospheric glow discharge with different gases. Their experimental results agree with the mentioned above considerations for selection of the working gas and plasma reactivation. Some lab experiments also indicate that the efficiency could be increased if using a mixture comprised of a working gas and a buffer gas. The working gas must have a low atomic mass and low ionization potential. The buffer gas could be an inert gas having a higher ionization potential and dielectric strength. Table 2.2 shows the Ionization potential and average atomic mass of some elements with low atomic mass that could be used as working gas. Note that some metallic vapors also could be used as working gases.

Table 2.2. Ionization potential and average atomic mass of selected elements

Element	H	He	Li	N	O	Ne	Na	Mg	K
Atomic Mass(amu)	1.0	4.0	6.9	14.0	16.0	20.2	23.0	24.3	39.1
Ioniz. (V) Potential	13.5	24.5	5.39	14.5	13.5	21.5	5.12	7.6	4.3

Another opportunity for increasing the efficiency of the SARG effect is apparent from analysis of the dynamic and static behavior of the CL node as a fundamental element of the underlying structure of the physical vacuum. (§2.9.1 Chapter 2 of BSM-SG). The applied AC field causes much larger CL node oscillations along the *xyz* axes than the oscillations responsible for a normal Zero-point energy. When a CL node is in a DC electrical field created by a large number of charged particles, its averaged central position is slightly displaced. If DC and AC fields are simultaneously applied, they will cause a CL node displacement plus dynamic oscillations of the EQSPM vector (see §2.9.4 of BSM-SG, Chapter 2). Since the CL node displacement has a limit, the combination of AC + DC field will cause a nonlinear effect, which does not exist at normal CL node oscillations (a normal state of the physical vacuum). This nonlinear effect will lead to more effective disturbance of the self-synchronization, which means an increased efficiency of the SARG effect. This predicted feature was verified experimentally, as discussed in the next chapter.

CHAPTER 3. Technical realization of the Heterodyne Resonance Mechanism. Gravito-inertial effect called a Stimulated Anomalous Reaction to Gravity (SARG).

3.1. Brief historical overview

The earlier research on massless propulsion using only EM fields is pioneered by Thomas Brown and Paul Biefeld and is known as the Biefeld-Brown effect [50]. Presently the study of massless propulsion is referred to as Electrogravity, Electrohydrodynamics, Electromagnetodynamics, Plasmomagnetodynamics, and so on. More often, all of these fields are referred to as Electrogravity. In the past 20 years, research in these fields has intensified in USA, Europe and Russia. Significant research about antigravity and propulsion has been done by individual researchers or private Labs. Among the serious researchers possessing a private Labs are G. Hathaway (Canada)[62], and J. L. Naudin (France) [63]. The provided so far research, however, is mostly experimental, since the physics is not well understood. When analyzing the effect from the point of view of contemporary physics, some contradictions to Newtonian laws of gravity and inertia appear. As a result, advances in these fields have been very slow and the results are not always repeatable. Some minor achievements were the result of extensive research efforts. Some of researchers derive empirical physical models, however, they are not fundamental and they appear quite dependent on the particular design. Such approach cannot lead to significant advancement. Despite all the efforts so far, the underlying physical mechanism is not well understood. Researchers and theoreticians have not envisioned the possibility of a gravito-inertial effect. The prior art research is summarized in the next Chapter 4. Here we will mention breafly some particular research in plasma field that has some connection with the envisioned Heterodyne Resonance Mechanism.

The plasma glow discharge was discovered by Nikola Tesla about 100 years ago. He first observed this interesting phenomenon and reported it at a special seminar of the Institute of Electrical Engineering, London, UK, 1892 [42]. The physics behind this phenomenon and some unusual effects have puzzled physicists up to the present day. Tesla obtained and investigated the plasma discharge of different gases in partial vacuum and also in air at normal pressure. The plasma discharge initially found applications in luminescent lamps and neon signs. After the II World War, serious research on plasma began in two main directions: high temperature plasmas and low temperature ones. The glow discharge phenomenon belongs to the second one. In 1961 B. Angerth et al. presented experimental results demonstrating an abnormal ionization mechanism [57]. They used a high potential electrical field in combination with a magnetic field. They claimed that a rapid ionization occurs when the velocity of plasma penetrating a neutral gas exceeds a critical value. Their experiments invoked a significant interest among researchers and theoreticians. G. Himmel and A. Piel also extended the applied technique to a helium-argon gas mixture [58].

3.2. Gravito-inertial phenomenon

Since the gravito-inertial phenomenon is predicted by the BSM-SG unified theory, the author first did extensive search on the published Electrogravity experiments. He analyzed them from the point of view of the BSM-SG concept and found common physical signatures. The unusual effects reported in some experiments were compared with the observed and documented effects that accompanied some UFO observations. Amongst them are the effects of reduced turbulence, EM interference, sound as a blowing wind, visual optical effects and identified optical spectrum. Apart from these, the author also analyzed the visual "mushroom-like" effects from the first few seconds of recorded movie clips of atmospheric tests of nuclear weapons. The fast development of the climbing column and the simultaneous appearance of multiple tornados in the first few seconds was carefully analyzed from the BSM-SG point of view. They clearly demonstrate a gravito-inertial effect (see manifesto "Prevent nuclear disaster" in [51]).

The author of this book came to the conclusion that some observed features in Electrodynamics, atmospheric nuclear tests and effects accompanying UFO observations share one comment signature – a gravito-inertial phenomenon. The author focused on EM activated plasma called glow discharge. Despite having been known for a long time, the physics of glow discharge has not been well understood in the prior art. The author did extensive Laboratory research in vacuum and in normal pressure. The theoretical prediction and experimental research led him to the

Chapter 3

discovery of an unique gravito-inertial effect that he called Stimulated Anomalous Reaction to Gravity (SARG) effect.

The SARG effect is a gravito-inertial phenomenon that could be activated by the Heterodyne Resonance Mechanism described in the previous chapter. The technical realization of this mechanism is called a Heterodyne Resonance method or simply a Heterodyne method.

3.3. Experiments for investigation of the Heterodyne Resonance Mechanism and demonstration of the SARG effect

The author provided a variety of experiments at reduced pressure (partial vacuum) and at normal air pressure in order to investigate the glow discharge. For this reason different cells were designed with the possibility to be filled with different gases and to operate at different vacuum levels. Most of the cells had 2 electrodes, some of them with axial electrode configurations, and others with radial configurations. Cells with a 3rd and a 4th grid electrode were also used.

Fig. 3.1 shows a cell with two activation and two grid electrodes. The grid electrodes were used for investigating the process at various test setups. The vacuum tubing 5 is connected to a vacuum pump with a pressure gage and a stopping valve. The tests are done at different vacuum levels.

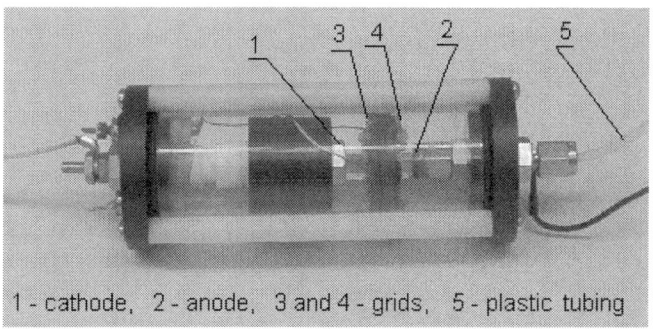

1 - cathode, 2 - anode, 3 and 4 - grids, 5 - plastic tubing

Fig 3.1. Vacuum cell with two activation and two grid electrodes

Fig. 3.2 shows one of the electrical circuits for study the glow discharge in vacuum. Block HV is a DC voltage converter. It charges a HV capacitor C_1 (12 uF) through a resistor R_1. The latter protects the HV output in a case of avalanche discharge in the

57

vacuum cell. The shown configuration of the capacitor C_2 the inductance L_1 and the vacuum cell allows invoking the Heterodyne Resonance Mechanism by tuning the resonance circuit C_3L_1. The oscillations are observed by an oscilloscope using the pickup coil L_2. The inductive coupling between L_1 and L_2 must be weak.

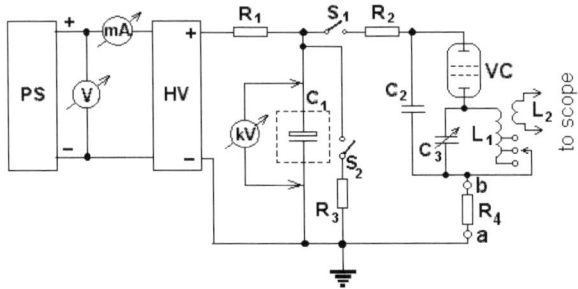

Fig. 3.2. Electrical circuit for study of glow discharge in vacuum
PS – low voltage power supply, HV - high voltage power supply, VC – vacuum cell.

The Heterodyne Mechanism can be identified by three major detectable signatures:
- very low power consumption
- emitted radiofrequency spectrum
- visual observation of a glow discharge around the cathode with some specific features discussed later.
- characteristic waveform shape measured by oscilloscope

One characteristic feature of the Heterodyne Mechanism is that the power consumption is quite small so HV block could be turned off and the discharge could operate from the charge of the capacitor C_{11} (12 uF) for one minute.

For activation of the glow discharge, the high voltage must be above some minimum value that depends on the type of gas, the pressure, and the distance between electrodes. The glow discharge mode is distinguished from the direct spark discharge. The DC power consumption of the glow discharge is extremely small - in order of uA. This is so because the ion-electron pairs oscillate reversibly not contributing to a DC current. The reversible motion requires a minimal free path length (without collisions), which depends on the gas pressure. The Heterodyne Mechanism, however, is not possible in a deep vacuum (without gas molecules) because ions and electrons are needed.

The oscillations observed by the pickup coil L_2 are comprised of consecutive bursts of high frequency sinusoids with exponentially falling amplitude. The average burst frequency depends on the pressure, the type of working gas and the value of the capacitor C_2 (see Fig. 3.2). The shape of a single burst of the high frequency oscillations when the resonance circuit C_3L_1 is not tuned is shown in Fig. 3.3. When the resonance circuit is tuned the burst duration is much larger and the envelope is different as shown later in Fig. 3.6.

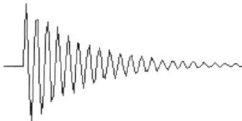

Fig. 3.3 Waveform of a single burst of high frequency oscillations

Tests provided with different gases under partial vacuum confirm the prediction that the Heterodyne mechanism will work better with gases from elements with lower atomic number. Tests with hydrogen show more stable oscillations with identification of electron spin flipping. This is apparent by the waveform of the oscillations and the measured spectrum. The activation and test circuit shown in Fig. 3.2 was used. For observing the Heterodyne mechanism, a precise tuning of C_3L_1 circuit is necessary.

Fig. 3.4.a. shows an EM spectrum emitted by glow discharge in a vacuum cell filled with H_2. The distance between the electrodes is 22 mm. Fig 3.4.b shows the EM spectrum when the same cell is filled with air. Both spectra are at a pressure of 8 mbar and an applied DC HV of 2kV. The measurement is made using a spectrum radiometer HP 8590L. The signal is picked up by a small antenna at a distance of 30 cm from the vacuum cell. The vertical scale is logarithmic. The spectrum from a cell with H_2 gas shows a peak at 2.3 MHz. The peak for a cell with air is at 2.4 MHz. The side lobes of the spectrum in the case of H_2 are much smaller.

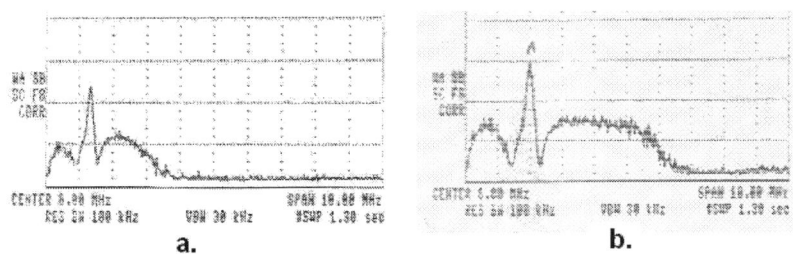

Fig. 3.4. EM spectrum from glow discharge at a pressure of 8 mbar and an applied voltage of 2 kV. a. cell filled with H2, b. – cell filled with air.

The experimental tests confirmed some signatures of the predicted Heterodyne Mechanism. While the EM spectrum from a cell filled with air is quite broadband, the spectrum from a cell filled with gas of low atomic number element such as hydrogen or helium is narrower. In the latter case, the observed waveform is more stable and clear. This is in agreement with the consideration (b) in section 2.2. that one should use a gas from elements with a lower atomic number.

For investigating the timing characteristics of the Heterodyne Mechanism, the HV circuit shown in Fig. 3.5 was used.

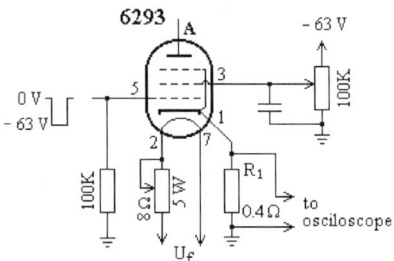

Fig. 3.5. HV circuit. This circuit replaces the tunable resistor R_3 in the circuit shown in Fig 3 and provides an opportunity for enabling/disabling the oscillating conditions and investigating the possibility of external synchronization of the burst rate.

The circuit diagram of Fig. 3.2 was modified in the following way. The resistor R_4 was removed and the circuit shown in Fig. 3.5 was connected between points *a* and *b*. The cathode–anode section of the tube 6293 replaces the resistor R_3 in the circuit shown in Fig. 3.2. This modification allows on/off switching of the oscillations by changing the potential of the first grid (G_1, pin 5). By controlling the input grid of 6293, it was possible to quickly turn on/off the high voltage applied to the VC in order to investigate the transient characteristics of the Heterodyne mechanism. The oscillations are enabled if G_1 is grounded and disabled when submitting a negative voltage to G_1 (about - 63V). The burst, however, could not be synchronized by on/off control of the G_1 grid when the rate is different from the intrinsic rate of the Heterodyne Mechanism process. The burst rate is quasiperiodical. Its average frequency can be regulated either by adjusting the cathode heating (very sharp tuning) or by regulating the potential of the first grid (G_1, pin 5 – sharp tuning) or the second grid (G_2, pin 3 - fine tuning). The voltage drop on the cathode resistor permits more accurate observation of the burst shape by oscilloscope.

Figure 3.6 shows single bursts of oscillations of vacuum cell with air at pressure of 12 mbar using the combined circuit of Fig. 3.2 and Fig. 3.5. At proper tuning of the resonance circuit L_1C_3 the burst frequency becomes longer and more homogeneous. The waveforms are measured by the pickup coil L_2 containing only 3 turns.

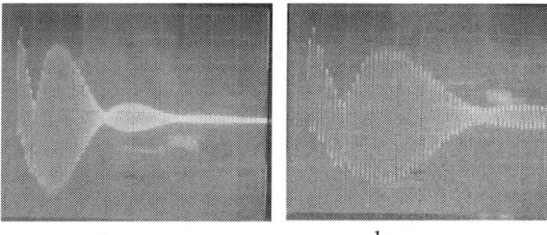

a.   b.
Figure 3.6. Single burst of oscillations from pickup coil L2 at two different sweep time bases.

Figure 3.7 shows a single burst waveform measured on the cathode resistor R_1 using the same test circuit. The grid G_1 potential was such that the tube 6293 did not have an anode current. However the oscillations are symmetrical. This means the direct current in this mode of operation is negligible. This is also evident from the negligible discharge of the capacitor bank C_1 when switch S_1 is open.

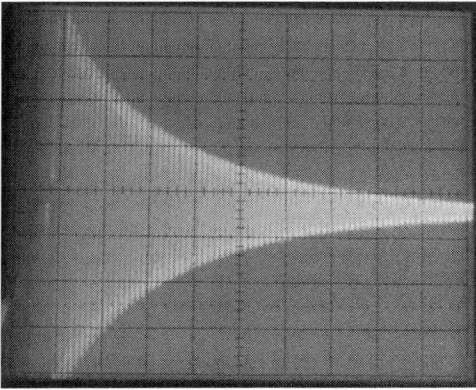

Fig. 3.7. Oscillations measured on the cathode resistor R_1.

From the study of the Heterodyne Mechanism in a vacuum cell, it was found that the burst phase could not be externally synchronized by on/off control of the G_1 grid of the radio tube 6293. This means that the negative resistance (a typical parameter used in oscillators) is an intrinsic feature of the oscillating glow discharge and it cannot be phase controlled externally. This is in agreement with the described physical mechanism involving a Quantum mechanical interaction between the oscillating electrons and one of the characteristic frequencies of the CL space, which in Earth gravitational field is equal to the Compton frequency.

The above conclusion is valid when the electron oscillations are invoked by a pulse method. It indicates that the phase of the oscillating electrons could not be suddenly synchronized by external means, because the electrons are engaged in a process of ZPE interaction with the physical vacuum. This conclusion is supported by another observed feature when using the circuit diagram in Fig. 3.2 without external switching. Despite the high voltage being kept constant, it was always observed that the burst rate is low at the beginning of the applied HV potential and increases gradually with time until reaching some limit. A similar effect was observed also in the circuit modification when tube 6293 was used and "on-off" switching was applied on the first grid G_1. The DC voltage level of G_1 was properly selected. The cycle of the control on/off voltage, however, must be longer than the intrinsic cycle of the Heterodyne Mechanism. The "on" state duty also must be properly adjusted. It was found that after a longer not operating time or initial start up, a longer time and higher HV is necessary for obtaining a stable self-oscillating mode.

Continuous mode of oscillations

Using the combined test circuit of Fig. 3.2 and Fig. 3.5, an additional feature for positive feedback was tested. For this purpose, a specially designed feedback was arranged using a device shown in Fig. 3.8. It is comprised of a metal cylinder 1, with an axial round electrode 2 having a proper diameter.

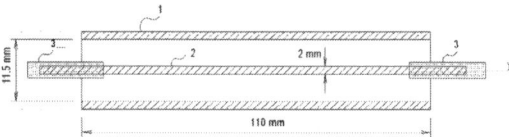

Fig. 3.8. Metal cylinder with radial electrode for LWs feedback
1- metal cylinder, 2 - electrode, 3 - isolator/metal layers

The signal from point c of Fig. 3.2 was conveyed by a 20 pF capacitor to the central electrode 2, while the metal cylinder 1 was connected to the first grid G_1 of the HV tube shown in Fig. 3.5. The purpose of the device shown in Fig. 3.8 is to transmit LWs in one direction - from the vacuum cell to the G_1, while not disturbing the Heterodyne process by a parasite reverse feedback from G_1. This is based on one characteristic feature of the LWs that they emit only from a convex surface in a radial direction but not from concave one. To prevent the edge effect both ends of the electrode 2 are covered by a few layers of metal and isolators. Metals layers are not connected between themselves. The LWs and their characteristic properties were discovered and studied by Nikola Tesla 110 years ago. He referenced them as Radial energy or waves. The diameter of central electrode 1 is selected to transmit LWs above some level. Insulators 3 prevent corona discharge from the sharp edges of 1 and 2. The observed oscillations from the Heterodyne process at proper resonance tuning and positive feedback is shown in Fig. 3.9.

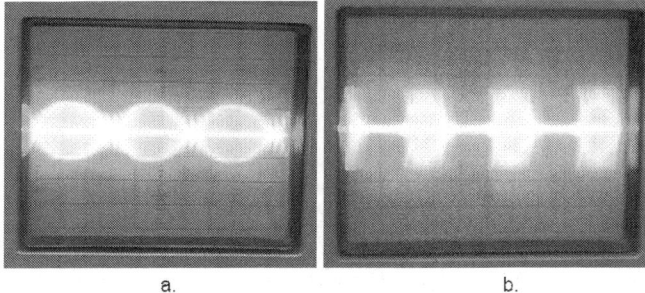

a.   b.
Fig. 3.9. Oscillations with a positive feedback

Fig. 3.9. a. and b. show the oscillations for cases of weak and strong LWs feedback. In the second case we see that the oscillations appear in patches despite the fact that the positive feedback is continuous. A possible explanation is that the strength of the feedback changes the step of the helical trajectory of the electron bound to a positive ion. These tests also confirm the previously mentioned feature that the duration of ion-electron pair oscillations is defined by the physical mechanism itself. In fact, the positive feedback changes the envelope shape of the oscillations. For a stronger feedback, the shape becomes closer to rectangular but the electron spin flip occurs faster and the process terminates until it is self-activated in the next burst of oscillations.

One interesting feature was detected while observing the peak of the spectrum with a higher resolution and simultaneous observation of the waveform on the scope. Fig. 3.10 shows the spectrum of a cell with H_2 at a pressure range of 12 (mbar) and a high voltage on C_1 in the range 1.5 – 2kV. The peaks are at 3.22 MHz and 3.67 MHz. Two peaks are always observed and they become equal only at fine tuning of the resonance circuit C_3L_1. In this case the waveform observed by a scope becomes longer with a slower exponential fall of the amplitudes.

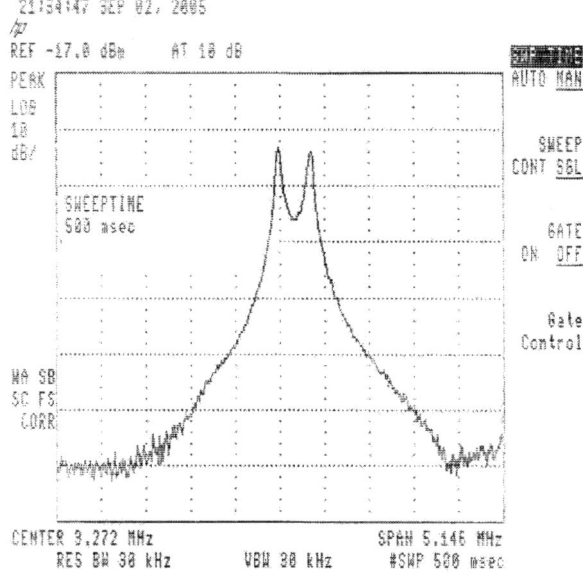

Fig. 3.10. High resolution EM spectrum from a cell with H_2 at pressure of 12 (mbar) and activating voltage between 1.5 and 2 kV. The double peak could be explained by an electron spin flipping.

Figure 3.11 shows 3 consecutive envelops of the HF oscillations measured by a digital scope. A small detuning of the C_3L_1 circuit leads to inequality of the peaks in the spectrum and inequality of neighboring envelopes of the waveform.

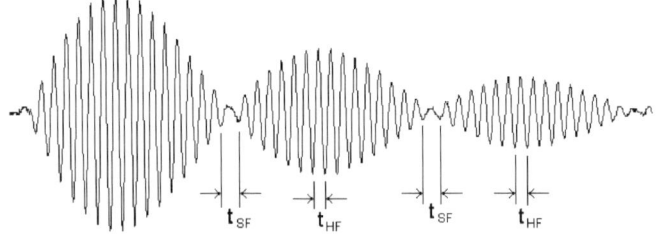

Fig. 3.11. Expanded section of a burst of oscillations measured by a digital scope. The difference between the periods t_{SF} and t_{HF} could be a result of electron spin flipping

At a lower pressure of 7 *mbar* the attenuation of the burst is slower, so its duration is longer. This is due to a longer free path between collisions. From the waveform in Fig. 3.6 it is apparent that:

$$t_{SF} > t_{HF} \qquad (3.1)$$

where: t_{HF} - is the high frequency period of oscillating ion-electron pairs, t_{SF} - is the time of spin flipping.

Both, the spectrum in Fig. 3.10 and the waveform in Fig. 3.11 indicate that an electron spin flipping occurs in the case of the invoked glow discharge. The condition (3.1) is an indication of spin flipping and it is in full agreement with the suggested electron model discussed in Chapter 2, §2.1. It is evident from Fig. 3.11 that the spin flipping does not occur on every HF cycle but about every 16 consecutive HF cycles. The number of these cycles depends on different factors: the type of gas, the pressure, the tuning of the L_1C_3 circuit etc. The change of the HF amplitude between flipping points indicates also that the momentum involved in the QM interaction between the electron and the physical vacuum (CL space) gradually changes and approaches a minimum value at the moment of electron spin flipping. All these features confirm the physics of the electron structure as suggested by BSM-SG and the predicted Heterodyne Resonance Mechanism.

Summarizing all of the test results discussed above, we conclude:

(1) When applying the ON-OFF mode using the Heterodyne method, by ON-OFF control of the activation HV, a minimum time is required between turning ON the HV supply and the appearance of the glow discharge.

(2) A stable glow discharge requires the period of the consecutive ON states to be larger than some minimum time defined by the process. Consequently, the ion-electron pairs require finite time for formation and they have finite time duration. The latter depends on the free path between collisions.

Fig. 3.12. shows two pictures of glow discharge in vacuum cells with different electrode arrangements. Visually, the glow discharge appears in two separate zones: one near the cathode and another near the anode. At a proper distance between the electrodes as in the case of Fig. 3.12.a, there is always a gap between the two zones. This gap is known as a Faraday gap. In the prior art of plasma physics explanations, it is assumed that this gap is needed for electrons to get enough velocity in order to cause ionization. However, this may explain only the column discharge near the anode.

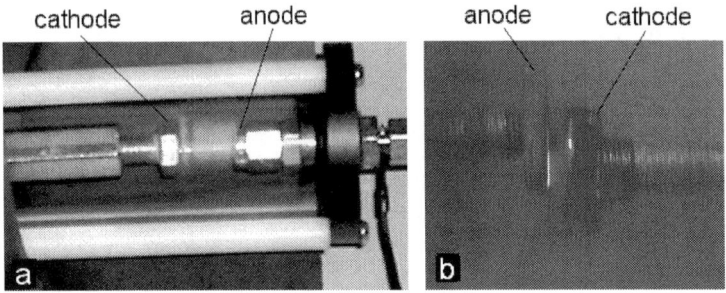

Fig. 3.12. Glow discharge in vacuum cells with different arrangements of the electrodes. The cathode is a steel screw with a hexagonal head. The dark envelope around the cathode in (b) is a glow discharge with a blue color (color pictures in http://ca.groups.yahoo.com/group/stoyan_sarg/

The curious feature of the glow discharge observed immediately at the cathode has not been so far theoretically explained. The glow discharge not only starts immediately from the cathode but even wraps around it. It envelops also at surface where the electrical field (between anode and cathode) must be negligible. Another effect is that when the vacuum cell is filled with air, the color of the glow discharge around the cathode is blue to violet, while the anode column has a pink color. When filled with a single gas, both columns have the same color. BSM-SG theory provides for the first time an explanation of the unusual features of the glow discharge around the cathode. The conclusion is: The ionization of the cathode glow discharge is caused by Longitudinal Waves discussed in Chapter 1, section 1.5. They emits in a radial direction from the electrode surface and preferably from edges with smaller radus. This type of ionization is more effective than ionization by accelerating electrons. The next experiment confirms this conclusion.

In order to be activated, the Heterodyne Mechanism needs an effective ionization mechanism. For testing possible ionization mechanisms, an experimental setup was made by using a Tesla coil and a glass globe filled with gas at a pressure of about 12 mbar. Instead of a spark gap, a pentode HV radio tube was used for the Tesla coil. This made it possible to obtain a HV output with a sinusoidal shape and to control its frequency.

Fig 3.13. shows a glass globe filled with Ne at a pressure of 12 mbar and placed above the top spherical terminal of a Tesla coil.

Fig. 3.13. Glass globe filled with Ne gas at pressure of 12 mbar placed above the spherical electrode of a Tesla coil (color picture in http://ca.groups.yahoo.com/group/stoyan_sarg/ - Photos)

The Tesla coil assures generation of Longitudinal waves (LWs) emitted by the spherical terminal. At a particular gas pressure, the observed glow discharge inside the glass globe is optimal at a particular frequency and magnitude of the LWs. The signature of this is the stable path and constant intensity of the observed discharge inside the globe. This means that the frequency must match the mean free path of the ion-electron pairs. The strength of the LWs is an important factor for the ionization. It was found also that there is some delay between turning HV on and the beginning of the glow discharge.

In Fig. 3.12.a, the distance between the two electrodes is almost equal to the radial size of the electrodes, while in Fig. 3.12.b the distance is smaller. The glow discharge in both cases has a different optical configuration. In **a.** there are two zones with a different color clearly separated by a gap, known as a Faraday gap. The location of the Heterodyne Mechanism is always near the cathode. One interesting feature is that, despite the DC electrical field being located between the two electrodes, the glow discharge wraps the cathode even behind the electrode head, where this field is quite weak. This is visible in both pictures **a.** and **b.**

Nikola Tesla first observed a glow discharge and reported some of its properties 100 years ago calling it "radial energy". He observed this effect also in normal air pressure. He used the term "radiant energy" because the observed stripes emanated at a right angle to the surface of the electrode and they preferred sharp edges. When the electrode lacks sharp edges, the stripes appear at higher potentials. These stripes are obviously caused by LWs discussed in section 1.5. In fact, the glow discharge that wraps the cathode is made of the tiny stripes demonstrated by Nikola Tesla [42]. The same stripes are also observed in the "plasma globe" if a non-sinusoidal AC HV is applied. Their first appearance is like "hair brushes" but later they become as "sharp threads". They obtain the latter shape due to some memory effect in the self-synchronization of CL space. <u>For obtaining a SARG effect in air, the "hair brushes" mode is required.</u>

Let us return again to the above-mentioned feature that the glow discharge appears also at the backside of the cathode, where the applied electrical field is negligible. Contemporary plasma physics cannot provide a reasonable explanation for this effect. From the BSM-SG point of view, the explanation is as follows: First, the LWs responsible for ionization are propagated at a right angle to the surface as discovered by Nikola Tesla. Second, in order to explain the wrapping of the electrode, we must assume that the LWs propagate faster than the electrical field. This is in agreement with the theoretical prediction in section 1.5 that Isotropic LWs propagate with a superluminal velocity over a short distance, while attenuating faster with the distance. Superluminal propagation now is reported in peer-reviewed journals [43].

Consequently, the observed effect is a signature of existing LWs propagating over a short distance with a superluminal velocity. The latter feature is discussed further in Chapter 4; where an additional unique application is based on a superluminal propagation of the LWs. LWs play an important role in the Heterodyne Mechanism because they cause the necessary ionization. They need to be emitted from an electrode that does not have sharp edges. Stronger LWs can be obtained by a more oblate electrode at higher potentials of the electric field. Sharp electrodes lead to corona discharge at lower voltages and the LWs are quite weak.

Fig. 3.14 illustrates a simple plasma thrust actuator for demonstration of the SARG effect. The activated plasma is observed as an envelope 7 that should not have a uniform thickness. The thrust force is in direction 8. At normal air pressure the actuator is supplied by a high voltage AC field in the order of 25 kV and adjustable frequency in the range of 2 – 5 kHz, combined with a HV DC field.

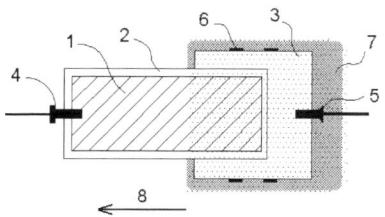

Fig. 3.14. Simple plasma actuator
1 - conductive body, 2 - isolation layer,
3 - teflon cap, 4 and 5 – electrodes,
6 – one or more floating potential electrodes

Fig. 3.15.a shows a simple version of the electrical block diagram supplying the necessary high voltage AC+DC for the plasma actuator. The DC high voltage consumption is negligible, so it is obtained by rectifying a fraction of the AC high voltage. The primary low voltage (not shown) is from a lead acid battery. This is according to the considerations of the T. W. Barrett theory [44].

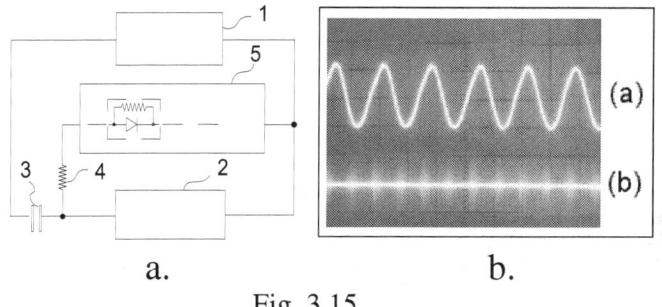

a.   b.

Fig. 3.15

a. Block diagram of the AC+DC HV power supply
1 – actuator, 2 HV AC power source with adjustable voltage and frequency, 3 – HV capacitor, 4 – resistor, 5 – rectifier group
b.   waveforms by oscilloscope. (a) – AC HV , (b)   HF signal generated by plasma

Fig. 3.15.b shows the waveform of the AC HV from the power source and the waveform of the high frequency signal generated by the plasma.

Fig. 3.16. shows a demonstration of the SARG effect as a continuous thrust force. Two simple plasma actuators are mounted on a horizontal bar supported at the middle on a small ball bearing.

Field Propulsion by Control of Gravity

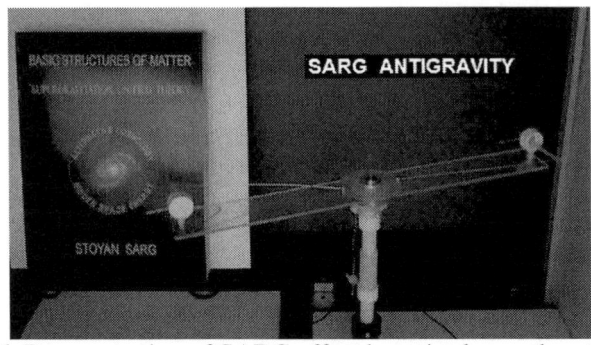

Fig. 3.16. Demonstration of SARG effect by pair plasma thrust actuators mounted on a horizontal bar (see the color image on the back cover and the videoclip in youtube.com)

The activating voltage is supplied by the ball bearing and one sliding contact. In this case only AC HV with amplitude of 25 kV and a frequency of about 3.5 KHz is applied. The SARG effect causes a rotation of the arm. The asymmetrical glow discharge of both actuators is visible. This experiment was demonstrated and reported at the 26 Annual conference of the Society for Scientific Exploration (SSE-2007, East Lansing, MI [9]). A videoclip was published on YouTube on May 16, 2007 [45]. The primary power supply was from a lead acid battery to avoid any connection to the power service since the plasma generates LWs that may pass through EM filters.

With this setup one interesting test was made. One of the conductors of the actuator was disconnected from the AC HV power supply and connected to a metal object of mass 2 kg at a distance of 2 m and away from any other equipment or ground. The actuator still generated a small thrust. Despite the circuit being open, the actuator still appeared to be getting power. In fact, the circuit was closed with respect to the LWs which propagate between two massive objects - the Led battery and the 2 kg mass. When the mass is much smaller, the LWs are not effectively transmitted and the actuator does not work.

In another test, the same 2 kg metal object was placed in a grounded Faraday cage. The same thrust was still observed. When the Faraday cage was left floating (not connected to ground) the thrust was diminished. This indicates that the LWs are not attenuated by a grounded Faraday cage but are attenuated by a floating one. When consecutive floating cages are made one inside another, the attenuation of the LWs is even stronger.

Chapter 3

When a very small corona discharge appears somewhere in the wires providing the AC HV, the thrust completely disappears.

All these tests indicate that <u>LWs accompany the SARG effect and play an important role in the process</u>. Unlike EM waves, the LWs pass through the Faraday cage when it is grounded. If the cage is not grounded it behaves like an antenna and the LWs are converted and emitted as EM waves. Unlike ordinary electricity, the LWs were partly conducted through isolators if they were dense materials. Glass, for example, is a very good isolator, but for LWs it is not a complete isolator. A less dense plastic would be a better isolator for LWs.

One important conclusion from the above considerations and from experiments is that any attempt to avoid interference by active ground separation or filtering does not work well in the case of LWs. Researchers should be aware of this. Otherwise, unwanted EM transients may damage sensitive equipment. In some cases, complete ground separation is necessary. In fact, Tesla Coil experimenters are aware of this problem. T. W. Barett provides some theoretical treatment about the issue of complete ground separation. [44].

In other experiment, a single actuator was hung on two wires. AC + DC HV was conducted in on/off mode to examine the thrust force. The combination of AC+DC field provided a larger thrust force. Videoclips in dark and light were posted on YouTube on October 5, 2007 [46,47].

In order to distinguish the propulsion effect from a possible ion wind, a single plasma actuator was enclosed in a transparent plastic cylinder as shown in Fig. 3.17.

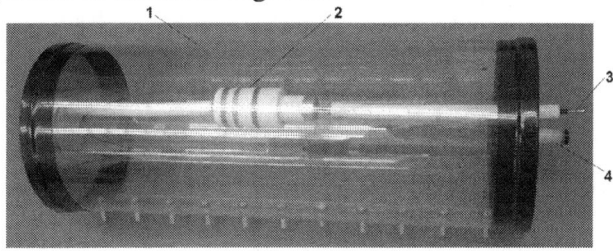

Fig. 3.17. Plasma actuator enclosed in a cylinder with gas at normal pressure. 1 – transparent cylinder, 2 – actuator, 3 – electrodes, 4 – valve

Periodically switching the AC+DC field on/off with the HV switch caused the actuator to begin to move as a pendulum. A short videoclip of the experiment was posted on YouTube on June 22,

2008 (Sarg propulsion effect) [48]. The propulsion effect in this case is smaller because of the weight of the cylinder with the enclosed air but it is still apparent. The direction of the thrust force always corresponds to direction 8 shown in Fig. 3.14. It is defined by the asymmetry of the plasma envelope. Since the applied AC frequency (in KHz range) depends on a number of design parameters as well as the type and pressure of gas, it must be adjusted until a noise like a blowing wind is heard. The gas inside the cylinder was air at normal pressure. With a different gas, such as Tetrafluoroethane from a dust cleaner for cleaning optics, the effect can disappear completely. When using Helium, a different actuator design with less sharp electrodes is needed, since the arc breakdown voltage for He is lower than for air. Note that the Heterodyne mechanism requires Longitudinal waves and they are best obtained at a high voltage potential.

To exclude a possible thrust from asymmetrical radiation due to emitted EM waves, the cylinder was further enclosed in a metal shield, not shown in the figure. The thrust (force field) was still present.

To exclude the possibility that the pendulum motion effect was caused by an ion wind inside the tube, the actuator was removed and a small fan was put inside. No pendulum motion was observed at any rate of on/off control of the fan.

One of the problems in activating a thrust actuator using a glow discharge is that it behaves as a capacitive load. Then the HV AC power supply must handle a large reactive power that does not contribute to the thrust but must be supplied by the HV AC power supply. This issue is addressed by a number of plasma researchers without a satisfactory solution [49,50]. The physics they derive, however, is based on empirical research and valid only for a particular design. None of the prior art researchers considered that the thrust force might be caused by some gravito-inertial effect. The empirical approach was not able to solve the problem with any efficiency. Some possible solutions to this problem are suggested in the next chapter of this book.

The Heterodyne Resonance Mechanism involves reversible motion of ion-electron pairs. However, their travel distance is restricted, and the free path length depends on the gas pressure. The colliding ions, atoms and molecules are sources of the emitted broad EM spectrum. By analyzing this spectrum, one may discover a signature of the Heterodyne Mechanism. In this respect, the emitted EM spectrum from different plasma-discharges was investigated.

Using a broadband spectrum radiometer HP 8590L, the EM spectrum emitted from the plasma actuators shown in Fig. 3.16 and Fig. 3.17 was measured. The EM signal was picked up by a broadband antenna connected to the sensitive input of the radiometer. Fig. 3.18. shows the EM spectrum at a distance of 2 m.

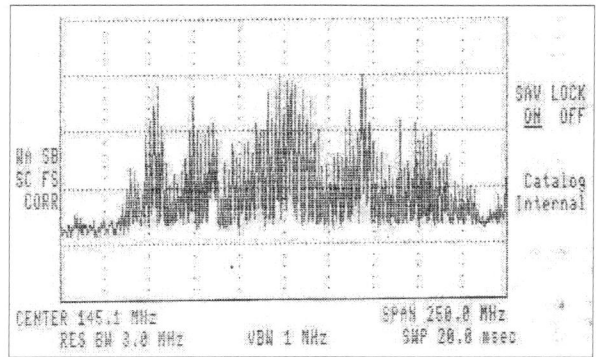

Fig 3.18. EM spectrum at 2 m distance from the experiment

The frequency of the activating HV AC field was 3.6 KHz, while the total power for generating the AC+DC field (included the reactive component) was about 50 W (not optimized power supply). The superimposed DC field was about 20KV.

The emitted spectrum from the EM activated plasma at air pressure covers a broader spectral range than at partial vacuum. The emitted spectrum for a gas with a low atomic number such as hydrogen or helium is much narrower. The broader spectrum range at air pressure is caused by the broader range of collision frequencies between the ion-electron pairs and other ions or neutrals. This means that the efficiency of the Heterodyne Mechanism must be lower at normal air pressure than when using a gas with lower atomic number at reduced pressure.

To test for possible particle or X-ray radiation from various plasma thrusters (supplied by a 40 W RF driver, which covers also the reactive power that returns to the driver) a military radiation detector IM-226/PDR-63 was used. No gamma or X-rays were detected at all. For the experiment in Fig. 3.16, some free charges (probably electrons) were detected at close proximity but the counts dropped quickly with distance and no radiation was detected beyond a distance of 70 cm. This confirms the existence of a spatial charge in the vicinity of the plasma thrusters. The spatial charge is one of the physical effects reported in a number of UFO observations.

3.4. Conclusions:

(1) Application of the SARG effect to a spacecraft propulsion system requires a specific design for the shape and geometry of the spacecraft, and a means of creating a partial or full plasma envelope around it. For maneuvering such a spacecraft, active control of the plasma envelope is needed.
(2) The SARG effect affects not only the mass of the actuator or spacecraft but also the mass of the surrounding air molecules. For this reason a reduced turbulence is observed in some experiments [49] and in some observations of UFO objects analyzed by the NASA researcher Paul Hill [30].
(3) The sound of a "blowing wind" is an audible signature of the SARG effect.
(4) The SARG effect is more efficient when using a mixture of two gases: a working gas with a low atomic mass and a buffer gas with a higher breakdown voltage. This will permit generation of stronger LWs.
(5) The emitted electrodes must be designed with a minimal curvature matching the LW requirements and without sharp edges. A parasite corona discharge may completely kill the SARG effect.
(6) The activated plasma provides a large reactive power that requires a specific approach for designing the HV AC source (more details about this are discussed in Chapter 4)
(7) Ordinary EM waves may not be suitable for communication between a spacecraft and the external world. Alternative communications with LWs could be used.
(8) Experiments with LWs must be done cautiously since they obviously have a biological effect.

The author announced the discovery of the SARG effect at the 27 Annual conference of the Society for Scientific Exploration (SSE), 2008 [10]. In 2009 the recorded talk was posted on the main website of the SSE and on YouTube.

The next chapter shows a patent application based on the discovered SARG effect. It was submitted to the Canadian Patent Office on 26 August 2008 [11]. The patent application also teaches the physics and the technical approach for creating a spacecraft shield to protect against micrometeorites.

CHAPTER 4. Method and apparatus for spacecraft propulsion with a field shield protection

Inventor: Stoyan Sargoytchev (Sarg)
Patent application No. 2,638,667 in CIPO, Canada
Filing date: 26 Aug 2008
Amendment filed on: 25 Nov 2008

ABSTRACT

The propulsion method is based on a gravito-inertial phenomenon predicted by the Basic Structures of Matter – Supergavitation Unified Theory (BSM-SG), the practical demonstration of which is called a Stimulated Anomalous Reaction to Gravity (SARG) effect. The SARG effect is a unidirectional change of the gravito-inertial mass of an object by modulation of the parameters of the physical vacuum. The suggested technique employs an asymmetrical envelope of EM activated neutral plasma. The result is a unique force field distinguished from reactive jet propulsion by the lack of throwing mass, and an effect of reduced gravito-inertial mass of the spacecraft and the surrounding gas molecules. This means less power for acceleration and less turbulence when moving in a planetary atmosphere. A small scale SARG effect is verified by laboratory experiments. A unique field shield protection against micrometeorites, also predicted by BSM-SG theory, can be achieved by emission of properly space and time correlated EM field packets and superluminal waves, known also as X-waves.

KEYWORDS: massless propulsion, space drive, field shield, X-waves, three-phase Tesla coil

BACKGROUND OF THE INVENTION

The present invention relates to a method and apparatus for creating a propulsion effect and a field shield protection to be used by a spacecraft preferably in deep space. The goal is to create a force field and a protective field shield without emission of mass particles, and using only electric, magnetic and electro-magnetic fields in a proper combination. In the prior art literature, such a propulsion system is referred to as a massless space drive or propulsion.

The main disadvantage of the massless propulsion methods described in the prior art is that they don't rely on an understandable physical mechanism. They are usually a product of experimentally discovered anomalous behavior without understandable physics. In most cases, the inventors or authors suggest explanations contradicting the known laws of physics or, at best, they propose empirical models without touching the contradictions with known physical laws. In both cases, the phenomenon could not be scaled and optimized for practical application, due to lack of understanding of the physical mechanism.

One of the massless propulsion methods, known as the Biefield-Brown effect was described initially in US patents 3,018,394 (1962) and 3,022,430 (1962). In the prior art literature, this effect is known also as Electrogravity. The effect is a weak propulsion force in the direction of the positive electrode of a capacitor-like actuator charged with a DC high voltage. The observable force field is possible only at high voltages above tens of thousands of volts. For practical applications, voltages in the order of hundreds of thousands and even millions are required, while a parasite arc discharge must be avoided. This puts severe constraints on the design of a spacecraft that must operate at different atmospheric pressures and in deep space. It is still disputable that the effect might be a result of an ion wind, since the polarity of the electrodes is constant.

Woodward et al. in US patent 6,098,924 describe an accelerator based on piezoelectric devices attached to resonant mechanical structures. The method lacks a physical explanation and only a small-scale effect is reported.

Podkletnov and Modanese [arXiv:physics/0108005v2, 2001] reported a small effect from an impulse gravity generator based on a charged YBaCuO superconductor. The major disadvantages are a cryogenically cooled environment (about -196 C), a high vacuum and a very small reported efficiency 1-2%. NASA sponsored expensive experiments in effort to replicate this experiment without success. The final attempt to replicate the experiment by G. Hathaway, with a 50 times higher accuracy and in consultation with Podkletnov is also without success. [G. Hathaway, Physica C, 385, 2003, p.488-500].

J. Reece Roth et al. in US patent 6,200,539B1 (2001) titled "Paraelectric Gas Flow Accelerator" describes an accelerator

Chapter 4

consisting of two sets of parallel metal strips on both sides of an insulating plate properly displaced. The strips of each set are connected together. When a high AC voltage is applied in the kilohertz frequency range, a specific glow discharge appears between the electrodes and one observes a weak acceleration effect relative to the surrounding gas atmosphere. Since the glowing plasma is obtained at normal atmospheric pressure, the method is called One Atmospheric Pressure Glow Discharge (OAPGD). The activated plasma emits a broad band RF spectrum in the range of 1 to 250 MHz. According to Roth, the observed small acceleration effect is a result of Lorentzian collisions of ions and electrons with the neutral molecules, atoms and radicals. This explanation is not satisfactory from a physical point of view because the plasma is excited at each half cycle of the AC field, so the electrode polarity alternates while the acceleration is unidirectional. For this reason, many researchers who investigate this type of accelerator express the idea that the effect is unknown. The effect exhibits also a small turbulence reduction for which there is not any physical explanation in the prior art.

Roth and other researchers consider that the thrust force is a kind of reaction of the surrounding. Thus, they do not envision possible operation in deep space and do not offer solutions for such applications. Since the physics of the observed phenomenon was not understood, they could not provide effective recommendations for the design of their accelerators and optimization of efficiency. There are two major disadvantages in trying to use Roth's OAPGD method in deep space. The first one is that, with the proposed method of plasma activation, the propulsion effect is very weak. The second one is that a small fraction of the power supplied to the accelerator, which behaves as a capacitor load, contributes to the force field, so the power efficiency is quite low. The passive network adapter suggested by Roth only slightly reduces the useless reactive power. Such an adaptor also could not be used at different environment pressures, which means working at different heights above the ground.

Another researcher and inventor S. Roy suggests a Wingless Electromagnetic Air Vehicle (WEAV) based on his research on the Dielectric Barrier Discharge. He does not go further than J. Roth about the physics of the phenomenon, proposing only empirical models for his particular model and does not propose means for use in deep space including a protective field shield against micrometeorites.

In the last 15 years, research in a field known as Electrohydrodynamics has intensified in the USA, Europe and Russia. Despite this, the possible existence of a gravito-inertial effect has not been envisioned in the prior art.

For a spacecraft moving with very high velocity, it becomes very necessary to have a protective field shield against micrometeorites in deep space and dust particles in an atmosphere. There is no provision for such kind of protection in the prior art.

SUMMARY OF THE INVENTION

It is an objective of the invention to propose a method and apparatus for propulsion of spacecraft with a field shield protection, in which the propulsion is a result of a unidirectional change of the gravito-inertial mass of the spacecraft, while the spacecraft is surrounded by a field shield that protects it from micrometeorites.

It is another objective of the invention to provide design considerations for the shape of the spacecraft with positions of the functional members of the propulsion system.

It is another objective of the invention to improve the prior art plasma thrust accelerators in order to be used in a spacecraft operating in a planetary atmosphere and in deep space.

It is another objective of the present invention to increase the efficiency of the prior art plasma thrust accelerators by applying simultaneously AC and DC high voltage fields.

It is another objective of the present invention to propose a circuit that prevents returning the reactive power from the capacitive type of plasma thrust actuator back to the AC high voltage power supply in order to increase the power efficiency of the force field.

The physics of the propulsion method and the field shield is provided by the Basic Structures of Matter – Supergravitation Unified theory (BSM-SG), published as a monograph by the author of this invention. The proposed propulsion method is based on the effect called Stimulated Anomalous Reaction to the Gravity (SARG). A small scale SARG effect was experimentally demonstrated in the laboratory. The protective field shield, the physics of which is also explainable by BSM-SG theory, relies on a combination of EM fields and superluminal waves, known as X-waves or Evanescent mode, emitted with properly selected parameters.

The invention, the theoretical bases and the experiments demonstrating its validity are described further below with reference to the accompanying drawings.

Chapter 4

BRIEF DESCRIPTION OF THE DRAWINGS

Fig. 1 illustrates the helical trace and the magnetic field of an electron bound to a moving single ionized atom, forming an ion-electron pair

Fig. 2 shows a simple thrust actuator for demonstration of the SARG effect.

Fig. 3 shows electrical means for activating the simple plasma thrust actuator demonstrating the SARG effect

Fig. 4 shows the waveform measured by an antenna at 1.5 m distance from the plasma actuator

Fig. 5 shows a dual-section plasma thrust actuator activated by AC High Voltage circuitry with increased power efficiency

Fig. 6 shows a basic electrode configuration for creation of a protective field shield

Fig. 7 shows the timing diagram of emitted signals for creation of a protective field shield

Fig. 8 shows the main functional blocks of a disc-shape spacecraft for a close range interplanetary flight

Fig. 9.a, b shows an overall shape of a spaceship for long-range travel with allocated positions of propulsion devices based on the SARG effect

Fig. 10. shows one preferable embodiment of a high voltage AC+DC circuit for the disc-shaped spacecraft

Fig. 11 shows a second preferred embodiment for generation of AC + DC high voltage for plasma activation in a disc-shaped spacecraft.

Fig. 12 shows the bottom electrode module for the second preferred embodiment

Fig. 13 shows the timing diagram of the generated DC + AC high voltage for the second preferred embodiment for a disc-shaped spacecraft.

DETAILED DESCRIPTION

The propulsion method is based on a gravito-inertial phenomenon predicted by the Basic Structures of Matter – Supergavitation Unified Theory (BSM-SG) developed and published as a monograph by the author of this invention. According to this phenomenon, the gravito-inertial mass of an object could be changed unidirectionally by proper modulation of the parameters of the physical vacuum. The experimental demonstration of this phenomenon is called the Stimulated Anomalous Reaction to Gravity (SARG) effect [11]. The application of the SARG effect in a spacecraft suggests the use of neutral plasma, partially or fully surrounding the spacecraft and activated by electromagnetic and electrical fields. The result is a unique force field distinguished from reactive propulsion by a lack of throwing mass, a reduced or eliminated reaction to acceleration, and reduced turbulence in atmosphere.

The suggested propulsion method has not been envisioned by Modern Physics because the concept of the physical vacuum adopted at the beginning of 20 century does not correspond to reality. After Albert Einstein developed his famous theory on General Relativity, he realized that the Ether is necessary. In his monograph Sidelights on Relativity (1921) [1] Einstein says:
"Recapitulating, we may say that according to the general theory of relativity, space is endowed with physical qualities; in this sense, therefore, there exists an ether. According to the general theory of relativity space without ether is unthinkable; for in such space there not only would be no propagation of light, but also no possibility of existence for standards of space and time. (measured-rods and clocks), nor therefore any space intervals in the physical sense".

The only argument of Einstein against the material Ether in 1921 is that physicists failed to build a working model based on Maxwell's assumption. Now it is known that the Michelson-Morley experiment is inconclusive due to a methodological error, namely: The Effect of Doppler shift is compensated by the effect of relativistic clock rate change. Both effects affect the wavelength so the expected interferometric fringe shift is nullified. (The effect of clock rate change was unknown at the time when the Michelson-Morley experiment was done). Michelson himself highly doubted the result, so he suggested other experiments with counter propagated light packets (Fig. 4 of Michelson-Morley paper [2], having a right intuition that the result will be quite different. Such an experiment was not funded during his lifetime. Original experiments based on interrupted counter propagated light packets were first realized by Prof. Stefan Marinov. In the period 1972-1982 he made three different laboratory experiments [3,4,5]. A number of other modern ether-drift experiments confirm our absolute motion through some existing space medium – Ether.

The BSM-SG theory [6,7,8,9,10] suggests that at the bottom level of all matter are two indestructible fundamental particles (FP) of different intrinsic matter with parameters related to the Planck scale of frequency and distance. In a pure empty space, these two particles interact by Supergravitational (SG) forces, which are distinguished from Newtonian gravitation in that they are proportional to the cube of the distance.

$$F_{SG} = G_O \frac{m_{01} m_{02}}{r^3} \qquad (1)$$

where: m_{01} and m_{02} – SG masses; r –distance; G_0 – SG constant that is different for FPs of the first and second type of intrinsic matter.

Under the SG law and pure geometrical restrictions, the two fundamental particles congregate in geometrical formations following a unique crystallization process (see BSM-SG theory Chapters 2 and 12). This process leads to crystallization of two prism-like sub-elementary particles with internal twisted structure, so they are called twisted prisms. They build both the underlying structure of space (physical vacuum) and the material structure of the elementary particles.

The underlying structure of space is called a Cosmic Lattice (CL) and it provides the known physical and quantum mechanical properties of the physical vacuum. The individual CL node is formed by 4 twisted prisms of the same type held together by (SG) forces. If considering as an isolated CL node, the 4 prisms are at mutual angles of ~ 109.50 corresponding to the axes in a tetrahedron. The CL space is formed from the two alternative types of CL nodes arranged like the atoms in a diamond. The SG forces between opposite kinds of CL nodes also may change the sign from their mutual distance, since they depend on the common super-high proper frequency. This gives the possibility for spatial gaps between the CL nodes and consequently a vibrational freedom. The same sub-elementary particles (twisted prisms) are also embedded in the material structures of the elementary particles. In the CL space environment (physical vacuum) the SG forces are strong at atomic scale distance, so they hold the protons and neutrons in the atomic nuclei. They correspond to the well known strong nuclear forces. One type of Van Der Waals force between closely spaced atoms and molecules is also a signature of the SG force. Another signature is the observed Casimir force between two closely spaced polished surfaces. SG forces are observed also in nanotechnology. The suggested physical model works quite well in all fields of Physics: Particle physics, Quantum Mechanics, Newtonian gravity and inertia, Special and General Relativity, atoms, molecules, and Cosmology. The existence of the physical substance of space, denoted in BSM-SG theory as a CL space, was confirmed by a number of modern light velocity experiments, which detected our absolute motion through space with a velocity vector of magnitude about 360 km/s (see Appendix1).

The flexible CL node has two axes of symmetry: one set of four axes denoted as abcd aligned with the axial axes of the twisted prisms and three orthogonal axes denoted as xyz axes. Both sets of axes define a unilateral tetrahedron. Each CL node made from one type of subelementary particles has four neighboring CL nodes made from the other type of subelementary particles. The two sets of axes of the neighboring CL nodes are aligned, while their oscillations only slightly affect their mutual distances and alignment. Investigating the dynamics of the CL node under SG law and more specifically the return forces along the two sets of axes, provided a new understanding of the relation between the electric, magnetic and EM fields on one side and the Newtonian gravity on the other. The oscillations along xyz axes involve SG forces which are thousands of times weaker than oscillations along the abcd axes. The EM field and light propagation involve mainly oscillations in a narrow angle along xyz axes, while Newtonian gravity appears as SG gravity propagated along the abcd axes.

The BSM-SG theory unveils the material structure of the stable elementary particles, such as the proton, neutron, electron and positron built of the same subelementary particles (twisted prisms) but arranged in helical structures with hierarchical order. The internal space volume of the helical structures contain internal space occupied by a lattice built of the same subelementary particles (twisted prisms but much denser than the CL structure, so the latter exerts a pressure on that volume. Using the unveiled structure of the electron and its quantum interaction with the CL space (Physics Essays, 16, No 2, 180-195, (2003)), the suggested model permitted expression of important physical constants by the CL space parameters (BSM-SG, Chapter 3, §3.13.3 – 3.13.4):

- **Static CL pressure**, PS - (a pressure exercised on the impenetrable volume of the elementary particle structure, defining the Newtonian mass of the elementary particle).

$$P_S = m_e c^2 / V_e = 1.3736 \times 10^{26} \quad (N/m^2) \qquad (2)$$

where: P_S - is the static CL pressure, m_e – mass of electron, c – speed of light, V_e – impenetrable volume of the electron structure

- **Dynamic CL pressure**, P_D - (related to the Zero Point Energy of Dynamic type and responsible for the electrical and magnetic fields and the quantum behavior of the elementary particles):

$$P_D = h v_c / (c S_e) = 2025.8 \quad (N/(m^2 Hz)) \qquad (3)$$

where: h – Planck constant, v_c - Compton frequency, S_e – impenetrable surface of the electron structure (cut torus)

- **Partial CL pressure**, P_P – related to the confined motion of the electron with one of its quantum velocities, in which the signature of the fundamental Fine Structure Constant α plays a role. The confine motion influences the motion of the atoms, molecules and solids as discussed in Chapter 10 of BSM-SG (from §10.4.1 to §10.4.6). From that analysis it was found that for electron velocity of $\upsilon = \alpha c$ the ratio between the Partial and Dynamic CL pressure is a function solely of the fine structure constant.

$$P_P / P_S = \alpha^2 (1 - \alpha^2)^{-\frac{1}{2}} \qquad (4)$$

Consequently, the fine structure constant plays important role for the quantum mechanical interactions between the electrons in confined motion and the physical vacuum. At the same time the fine structure constant is embedded in the structure of the electron (Chapter 3 of BSM-SG) and its signature is persistent in the quantum electron orbits (see §7.5 - §7.8 of BSM-SG, Chapter 7).

The Dynamic CL pressure is related to the Zero Point Energy envisioned by Quantum Mechanics. The Static CL pressure is related to a hidden Zero Point Energy, envisioned by the BSM-SG theory. For this pressure the Einstein Equation $E = mc^2$ is valid. Using the Static CL pressure, the mass equation (5) of a stable elementary particle is derived (BSM-SG, Chapter 3).

$$m = \frac{4h\nu_C^4(1-\alpha^2)}{\pi\alpha^2 c^5} V_{ep} \qquad (5)$$

where: V_{ep} – is the impenetrable volume of the elementary particle, h – is the Planck constant

The Complex CL node dynamics is characterized by two identifiable cycles – one with a proper resonance frequency $\nu_R = 1.093 \times 10^{29}$ Hz (defining light velocity as one cycle per one CL node distance) and the Compton time with a frequency $\nu_C = 1.236 \times 10^{20}$ Hz, defining the permittivity and permeability of the physical vacuum (BSM-SG, Chapter 2). The Compton cycle involves a number of whole resonance cycles.

Since the neighboring nodes are interconnected by SG forces, the following unveiled features of the CL node are quite important for understanding the properties of the physical vacuum:

• SG forces are based on frequency higher than the proper resonance frequency of the CL node, while their propagation through CL structure as Newtonian gravity depends on the mutual phases of the oscillating CL nodes.

• The mutual interaction between the oscillating CL nodes causes a self-synchronization effect at the Compton frequency that is a second proper frequency of the CL node (defined by a full spatial cycle of the SPM vector)

• The effect of self-synchronization appears as permanently existing and recombining Zeropoint waves. They are responsible for the constant value of ε_0 and μ_0 defining the constant light velocity according to the formulae

$$c = (\varepsilon_0 \mu_0)^{-0.5} \qquad (6)$$

In Chapter 10 of BSM-SG, it was shown that the inertia of a solid object is related to the integral inertial momentum of displaced and folded CL nodes, which is expressed by the force moment vector, E_{IFM}. This vector, defining the inertial properties, is able to describe any kind of motion: uniform, rotational or accelerated. For a single particle with mass m it is:

$$E_{IFM} = c\alpha m\upsilon \qquad (7)$$

where: c – speed of light, α - fine structure constant, m – particle mass, υ - velocity.

Chapter 4

Equation (6) shows that the E_{IFM} vector will get a directional velocity if c is affected asymmetrically by selfsynchronization disturbance.

The validity of the mass Equation (5) and the inertial property Equation (7) propagates to atoms, molecules and also to a solid object. The latter is regarded as an integral entity of stable elementary particles.

Conclusion: Asymmetrical disturbance of the selfsynchronization around an elementary particle, a neutral atom, a molecule or a solid object will cause a change of its gravito-inertial mass according to Eq. (5) and a unidirectional non-inertial displacement according to Eq. (7). From both Equations it is evident that the common parameter, c – speed of light, should be affected if achieving an interaction with the Compton frequency $v_C = 1.236 \times 10^{20}$ Hz, which is one of the basic parameters of the physical vacuum.

Now the question is how to access this super-high frequency. The answer comes from the unveiled structure and oscillating properties of the electron. It is shown in BSM-SG, Chapter 3 (also in Physics Essays, 16, No. 2, 180-195, (2003)) that the electron possesses a material structure of a cut toroid as a single turn coil with a radius – the known Compton radius and a small helical step responsible for its anomalous magnetic moment. Then the electron's material structure appears as a three body oscillating system exhibiting a screw-like motion with oscillation property characterized by two proper frequencies. The first proper frequency of the electron is the known Compton frequency, while according to BSM-SG theory the CL node possesses the same Compton frequency. The second electron proper frequency is 3 times the Compton frequency and plays the role of the electron's spin. In such aspect, the moving and oscillating electron has preferred screw-like motion velocities defined by its Quantum Mechanical interaction with the physical vacuum. It is found that the strongest QM interaction is at electron velocity of $V = \alpha c = 2,187 \, (km/s)$ corresponding to energy of 13.6 eV. At this velocity the phase of the moving and oscillating electron matches the CL node phase propagated with the speed of light. Other Quantum velocities are $(\alpha c/2)$ - corresponding to energy 3.41 eV, $(\alpha c/4)$ corresponding to energy of 1.51 eV and so on. The oscillating model of the electron explains quite well all known properties of the electron including its anomalous magnetic moment, spin, gyromagnetic factor and the way it creates quantum orbits in atoms and molecules.

For accessing the Compton frequency a technical approach called the Heterodyne method is suggested. According to this method, the Super-high Compton frequency of the CL node $v_C = 1.236 \times 10^{20}$ Hz, which is a basic parameter of the underlying structure of the Physical vacuum, can be reached by the oscillating electrons, each one bound to a single ionized atom.

The physical mechanism of the Heterodyne method is illustrated in Fig. 1, where 1 – is the trajectory of the single ionized atom, 2 – is the helical trajectory of the bound electron, 3 is the magnetic field line of the electron moving in a helical trajectory, 4a and 4b are electrodes on which AC high voltage is applied. Considering a moving ion with a trajectory 1, the bound electron will make a helical trace 2. If the positive ion motion is reversible, the bound electron will also make a reversible helical motion. Since the helical step of the electron's structure mentioned above is much smaller than the electron's Compton radius, the confined motion velocity of the electron moving on the helix 2 will be much greater than the ion velocity. This allows the electron to move in a helical trajectory with one of its quantum velocities corresponding to energies of 13. 6 eV or 3.41 eV, while the velocity of the ion can be much smaller. It is well known that the magnetic moment of the electron is 658 times greater than the magnetic moment of the proton and 981 times greater than the magnetic moment of the neutron. Then the magnetic field of the bound system of the ion-electron pair will be predominated by the magnetic field created only by the electron. Further, the magnetic field from the electron moving in a helical path is many times stronger than if moving with the same velocity in a straight line. Additionally, the magnetic fields of the neighboring ion-electron pairs interact constructively. As a result, a large number of ion-electron pairs move in a cluster.

The described mechanism is achievable for an actuator containing two electrodes separated by a proper gap and operated in a gaseous atmosphere with conditions for ionization. Then one may create an alternative electrical field by applying an AC high voltage in the accessible RF range so that the ion moves reversibly, while the bound electron moves in a helix with one of its most probable quantum velocity corresponding to energies of 13. 6 eV or 3.41 eV. The reversible motion of the electron under the applied AC field causes a flipping of the electron spin. From the point of view of the unveiled electron structure [8] the spin flipping is a change of the

Chapter 4

phase of the oscillating electron at the Compton frequency by 180 deg relative to the Compton frequency of the CL nodes. At this particular moment, strong energy interactions take place between the electron and the physical vacuum.

A practical realization of such a process is the creation of neutral plasma around a properly design actuator. The physical process is characterized by the following consecutive phases:
- ionization of neutral atoms or molecules
- ions get acceleration
- build up of ion-electron pairs, each one formed by a single-charge ionized atom and a free electron
- the ion-electron pairs are initially accelerated by the applied electrical field acting initially on the positive ions and after that guided by the common magnetic field created by the bound electrons
- the acceleration and motion of the electron-ion pairs is reversible in every half cycle of the applied AC electrical field

The following considerations are dictated from the time duration and efficiency of such type of plasma:
- The reversible motion of the ion-electron pairs can be disturbed by collision, so the process needs reactivation
- The process is more effective for ions that contain lower number of protons and neutrons since the ratio of the electron to proton (neutron) magnetic moment ratio is greater.
- The frequency of the applied AC field must assure that the ions are trapped between the electrodes.
- An external magnetic field properly oriented in respect to the magnetic field of bound electrons might increase the efficiency.

The Heterodyne method is achievable in EM activated neutral plasma excited by AC high voltage or other means. A signature of positive ion-electron pairs is found in the analysis of a large number of experiments involving EM activated neutral plasma. It is apparent in experiments using different gases and at different pressure ranges – from a vacuum to a normal atmospheric pressure and even above normal pressure. In many published experiments the researchers point out that the average electron velocity is in the range of 3-10 eV, which is much lower than the expected one and they did not have an adequate explanation for this. In fact this is in a good agreement with the predictions of the Heterodyne Resonance effect, since the estimated average electron velocity in fact is an average value of the electron velocities involved in the ion-electron pairs (energies of 13.6 eV and 3.41 eV) and some free electrons. The formation of ion-electron pairs must be done frequently due to their depletion from collisions with neutral molecules and other ions. Thus, at normal and reduced atmospheric pressure it is convenient to apply an AC HV field, which can be in the kHz frequency range.

Practically, the Heterodyne method can operate in a wide range of pressures from a few tors to atmospheric and even above atmospheric pressure. The formation of ion-electron pair clusters helps to increase the effective free path. Nevertheless the effect strength is reduced in the normal atmospheric pressure due to losses from collisions and presence of negative oxygen ions. The efficiency is strongly dependent on the average free path between collisions. The collisions contribute to the significant fraction of the broadband high frequency spectrum, which is unwanted EMI noise. Only single ionized positive ions can participate in the working ion-electron pairs. Some negative ions significantly disturb the process. The selection of a working gas is also important, as this is evident from the considerations mentioned above. Massines et al. [12] investigated an atmospheric glow discharge with different gases. Their experimental results completely agree with the considerations mentioned above for selection of the working gas and plasma activation. Some lab experiments also indicate that the efficiency could be increased by using a mixture of a working gas with a lower ionization potential and a buffer gas with higher ionization potential.

One additional important feature for increasing the efficiency of the SARG effect was predicted by the inventor when analyzing the dynamical and static behavior of the CL node as an element of the physical vacuum structure according to BSM-SG (Chapter 2). The applied AC field causes much larger oscillations of the CL node along its xyz axes than the oscillations responsible for a normal Zero-point energy. When a CL node is in a DC electrical field created by a large number of charge particles, its averaged central position is slightly displaced. If DC and AC fields are simultaneously applied, they cause a CL node displacement plus dynamic oscillations. Since the displacement has a limit, the combination of strong AC + DC field causes a nonlinear effect that does not exist in a normal CL node oscillation. The nonlinear effect leads to increased efficiency of the SARG effect, which means the synchronization disturbance that leads to creation of the force field mentioned above is more effective. The inventor experimentally verified this feature, not envisioned so far in the prior art research.

Fig. 2 shows a simple plasma thrust actuator for demonstration of the SARG effect, comprised of a conductive body 5 covered with an isolation layer 6, a teflon cap 7, a first electrode 8, a second electrode 9, and floating electrodes 10. The activated plasma is observed as an envelope 11 that may not have a uniform thickness. The thrust force is in direction 12. At normal air pressure, the actuator is supplied by a high voltage AC field in the order of 25 kV and adjustable frequency in the range of 2 – 5 kHz, combined with a DC field. In order to eliminate the possible effect of ion wind, the actuator was enclosed in a transparent plastic cylinder and hung on the two thin wires that connect the electrodes 8 and 9 to the HV AC+DC power supply. In order to exclude a possible effect from asymmetrical radiation pressure due to emitted FM radiation, the cylinder was also enclosed in a metal shield, not shown in the figure. The thrust force was always in the direction 12. It is defined by the asymmetry of the plasma envelope 11. Since the AC frequency depends on a number of design parameters including the type of the gas and its pressure, it must be adjusted until hearing a specific noise like a wind blowing.

Fig. 3 shows the block diagram of the electrical circuit used for testing the plasma thrust actuator illustrated in Fig. 2. It is comprised of an AC high voltage generator 14, a capacitor 15, a resistor 16, and a high voltage rectifier 17 containing a number of diodes shunted with high ohm resistors and connected in series. By setting a proper value of the resistor 15, this circuit assures that the necessary DC high voltage is always proportional to the magnitude of the AC high voltage.

Fig. 4 shows the signal measured by antenna at 1.5 m distance from the plasma thrust actuator, where 18 is the AC HV frequency signal and 19 is the packet signal containing a RF frequency in a range from a few MHz to about 200 MHz. The RF signal is a result of collisions between ion-electron clusters with the free ions, atoms and molecules. For uniform plasma discharge, the RF packet has a duration much smaller than half of the sinusoid. The experiment is also accompanied by a clear audible signal of a pure sinusoidal component corresponding to the AC frequency and a noise like a wind blowing through an obstacle. The blowing wind-like sound is from the collisions between the reversible moving ion-electron clusters and the air molecules.

The observed SARG effect in this laboratory experiment is weak due to the following reasons, predicted theoretically and confirmed experimentally:
- The atmospheric air is not the optimal gas mixture for the Heterodyne method. Lower atomic mass gas as Helium is much more effective, but it has a lower dielectric strength. The solution is to use lower atomic mass gas with a buffer gas having a higher dielectric strength
- The SARG effect has a nonlinear dependence on the applied electrical field and has a bottom threshold limit. The applied field for the experimented test bed is about 20 – 25 kV. Higher voltages require special laboratory environments.
- From the analysis of the CL node dynamics (BSM-SG, Chapter 2) in case of an applied electrical field it becomes apparent that the force field should be proportional to the HV in the order of U^2 to U^3. Different researchers of plasma actuators with AC activation also report a similar range.
- From an energetic point of view, the plasma activation by a sinusoidal AC high voltage is highly inefficient, because the actuator behaves as a capacitor load with a negligible DC current discharge. As a result, a large reactive power returns to the power supply and dissipates as heat.

The theoretical predictions of the Heterodyne method, verified by experiments, indicates that instead of a full AC sinusoid only the initial activation slop of the AC high voltage with finite time duration is necessary for creation of oscillating ion-electron pairs. The finite time duration depends on the working pressure. While the AC field provides this condition every half cycle, it is obvious that the unwanted reactive power could be more effectively rejected if a feedback cut-off is implemented at exact phase point of the activating AC voltage. One possible way to provide such cut-off is an AC HV source based on a Tesla coil, which incorporates a spark gap in the primary coil. The spark gap plays the role of a plasma switch that may cut-off the unwanted parasite feedback at the required phase point of the AC cycle. Another option is using a Marx-bank circuitry. Other AC HV circuits based on tyratrons or tyristors also could be used with some compromise on the power efficiency.

Fig. 5 illustrates another option of plasma actuator comprised of a conductive body 20, isolator caps 21 and 22, end side electrodes 23a and 23b, and a medium electrode 24 connected to the conductive body 20, which may optionally have an isolating cover like the actuator in Fig. 2. The actuator has two opposite plasma sections between the two opposite electrodes 23a and 23b and the medium electrode 24. In this case the middle of a HV transformer 25 is connected to the electrode 24. The diodes 26 and 27 assure alternative operation of the opposite plasma sections, while the reactive energy from each one section will be transferred to the other one by the circuits 28 and 29 consisting of an inductance, capacitor and diode connected in series. This transfer must occur at the proper point of the sinusoid of the activating AC voltage, so it puts more severe constraints to the activating AC frequency, which depends on the working gas and the atmospheric pressure. The means of assuring simultaneous DC high voltages to both sections are not shown in Fig. 5, but they can be obtained in a similar way as shown in Fig 3. This type of plasma actuator must assure higher power efficiency in comparison with the actuator shown in Fig. 2 supplied by the circuit shown in Fig. 3.

The creation of a protective field shield around a spacecraft is envisioned from the analysis of the dynamical behavior of the CL nodes in applied electrical magnetic and EM fields, according to the BSM-SG theory. The CL space model of the physical vacuum predicts also the existence of compression-like waves, different from the known EM waves. A number of theoreticians have derived such waves by using the original Maxwell equation based on quaternions [13,14]. Such waves are also confirmed experimentally and they are known in the prior art as scalar or longitudinal waves [15]. They are able to carry much greater energy than the ordinary EM waves. The pioneer in generating longitudinal waves is Nikola Tesla, who provided the means for their generation in tens of patents and lectures in the period from 1893-1913 [16]. The longitudinal waves have a clear physical explanation from the point of view of BSM-SG theory: The EM waves involve oscillations along xyz axes of the CL nodes, while the longitudinal waves involve oscillations along abcd axes. The stiffness along *abcd* axes is thousands of times stronger than the stiffness along the xyz axes, so the longitudinal waves may carry much larger momentum than the EM waves. One additional feature of the longitudinal waves, also envisioned by BSM-SG theory, is that they may propagate much faster than the speed of light if the CL nodes in their path are synchronized. From the analysis in BSM-SG theory (Chapter 2, section 2.10.4) it becomes evident that within the photon wavetrain the CL nodes are synchronized. This means that within the wavetrain one can propagate information and energy much faster than the velocity of light. This is confirmed by some experiments called "quantum teleportation" which demonstrate transfer of information much faster that the speed of light. The analysis of the methods for producing femtosecond laser pulses from the point of view of the BSM-SG theory also unveils a superluminal effect of squeezing the extending beam path with a finite length into a strong femtosecond pulse. In the case of EM wave propagation, the CL node dynamics is similar to the photon wavetrain with the difference that there is not a transverse boundary limit as in the wavetrain of the photon.

The experiments of N. A. Kozyrev, provided in the period 1977-1982 [17,18] and later repeated by other Russian scientists [19] are known in Russian literature as the Kozyrev effect. In fact, the Kozyrev experiments exhibit two major phenomena. One is a reduction of the gravitational mass of a solid object by a small fraction after some mechanical (vibrational) or electrical treatment. He also found that the weight restoration takes a finite time in the order of minutes. The other phenomenon found by Kozyrev is a determination of the present positions of some distant stars by detection of non-EM waves from them traveling at a billion times the speed of light. This superluminal velocity was detected also by Gregory Hodowanec during a moon eclipse [20]. Kozyrev found also that superluminal non-EM waves exhibit a gravitational effect – they affect the position of a supersensitive weight balance used as a detector [17]. Finally a number of experiments published in recent years in peer-reviewed journals demonstrate a superluminal propagation of waves known as X-waves in a closed ranged field of a few wavelengths. D. Mugnai, A. Ranfagni and R. Ruggery [21] demonstrated superluminal propagation of microwave packets at 8.6 GHz (wavelength of 3.48 cm) up to a distance of about 1 m with average superluminal velocity greater than light velocity by 5.3%. The authors expressed the idea that there is not a theoretical limit for superluminal propagation at longer wavelengths

Based on the prior art experiments and observation of some phenomena analyzed from the point of view of BSM-SG, I came to the following conclusion:
It is possible to create an artificial boundary on the propagated EM wave with unique properties of a protective field shield. I suggest the following technique:
- Emission of EM wave packets of stable frequency in the RF or microwave spectral range

- Emission of a delayed strong short EM pulse with a large dU/dt and the same aperiodic frequency but with higher order odd harmonics with the same phase
- The EM wave packet and the strong EM pulse must be emitted from one and the same location and from a common or a separate circular dipole antenna with length equal to one or multiple wavelengths.

Fig. 6 illustrates the timing diagram requirements, where 30 is the EM wave packet with a duration t_{EM} and 31 is the strong EM pulse emitted at a time after the front end of the wave packet 33. The shown waveforms must be repeated with a period larger than t_{EM}. When emitted by a properly selected means during the EM wavepacket, the strong EM pulse will propagate at the close range field with a velocity exceeding the light velocity, an effect called superluminal propagation.

According to the above-mentioned conclusions about the superluminal propagation of a strong EM pulse along the path of the EM wave packet, it is obvious that the delayed strong pulse will propagate along the path of the EM wave, but will override it until the front end of that packet is reached. If the propagation time of the EM wave packet to this point is t, the time propagation of the superluminal pulse is $(t - t_d)$. From this moment, the conditions for the superluminal propagation of the strong pulse become suddenly different and a kind of reflection effect will occur. This will also affect the further propagation of the EM wave packet, and its energy will be deposited at this moment in a thin layer, forming something like a compressed zero point energy of the physical vacuum. If assuming for simplicity that the EM wave packet and the strong pulse are emitted by a spherical emitter, the aforementioned layer will correspond to a spherical surface with a radius R according to Eq. (8)

$$R = \frac{V_x t_d c}{V_x - c} \tag{8}$$

where: V_x is the superluminal velocity of the strong EM pulse in a field distance of a few wavelengths of the EM wave packet frequency, c – is the light velocity

For the described case of emission of a strong EM pulse during the emission of an EM wave package, the value of V_x must be determined experimentally, since it may appear much larger than the estimated one in the experiment of D. Mugnai et al [21].

The deposition of the whole energy of the EM wave packet on the spherical surface with a radius R and small thickness means that the self-synchronization of this spherical surface will be disturbed or rearranged. It was theoretically predicted by BSM-SG theory and confirmed by some observed phenomena that the disturbed or rearranged self-synchronization needs a finite time for self-restoration. If the time between the emitted EM wave packets is shorter than the self-restoration time, many wavepackets will dissipate their energies until some energy balance occurs. The disturbed self-synchronization of the CL nodes on this spherical surface will affect the gravito-inertial mass of the dust particles, while the dissipated energy will create some kind of protective field shield against micrometeorites. According to the BSM-SG theory (Chapter 3, section 3.12.2.A) the disturbed self-synchronization will affect the conditions for the quantum orbits and, consequently, it will cause a weakening of the atomic and molecular bonds in the solids. Therefore the micrometeorites moving with large velocities might be disintegrated into smaller fractions.

When the field shield is created at conditions of normal Earth atmospheric pressure, part of the emitted EM energy will dissipate within the volume of the sphere due to partial ionization of the gas molecules. Therefore, the protective field shield will be more efficient in a highly rarefied atmosphere. The analysis predicts the possibility for creation of a field shield over a spacecraft moving in deep space, since the gas mixture released by the spacecraft for the necessary plasma may provide the necessary conditions for a field shield up to some radius. Further, the described effect might be useful for more than spacecraft protection. One may speculate that if applied on a larger scale on a planet with or without a rarefied atmosphere, the protective field might form a spherical dome inside of which an artificial atmosphere could be created. The field shield would serve as a stop boundary for the artificial atmosphere in order that it could not escape into deep space. In such a case, an artificial colony could be created, for example, on the Moon or on Mars.

Fig. 7 shows an example of the electrode configuration for a means of assuring a protective field shield, where 32 is a circular dipole antenna for emission of the EM wave packets, 33 is a sectored ring antenna for emission of the strong EM pulse, and 34a and 34b are spherical electrodes with a common connection serving as a virtual ground for the sectored ring antenna 33. Since the emitters do not have a spherical emission diagram, the protective field shield will still be spherical but will be stronger at radial directions where the dipole emission is stronger. This is illustrated by the density of the dashed line showing the protective field shield 36, while 35 shows the emission diagram of the dipole antenna 32. The dipole antenna 32 and the ring antenna 33 have the same diameter, and their circumferences are equal to the wave packet wavelength. The virtual grounds for both antennas are completely isolated. This measure not only protects them from mutual interference but also allows operation with larger energy according to the Nonlinear Oscillator-Circuit Theory published by T. W. Barrett [22]. The ring antenna 33 is cut at least in one place, so it contains one or more gaps. This measure prevents it from behaving as a short turn when the wavepacket is emitted by the dipole antenna 32. However, when emitting a strong EM pulse, the antenna 33 behaves as a single ring, since the gaps are of such a length that they will be crossed by sparks appearing during the strong pulse. The ring antenna 33 can be used also for activation of the neutral plasma between it and the electrodes 34a and 34b for creation of the SARG effect. Such a twofold function of the antenna 33 allows a better relation between the necessary propulsion force field and the protective field shield. For this purpose, however, a more complex electrode configuration is required, as will be shown below.

Fig. 8 shows the shape of a spacecraft for interplanetary flight with the external functional elements of its propulsion system, where 37 is a spacecraft body with a thick isolation layer for withstanding high voltage potentials, 32 is a dipole antenna, 33 is a ring antenna, 38 is an upper spherical electrode, 39 is a set of three bottom spherical electrodes at 120 deg, 40 is a set of isolated electrodes, and 41 are set of portholes for preactivated plasma. The purpose and functions of the dipole antenna 32 and the ring antenna 33 are previously described by the help of Fig. 6 and Fig. 7. The oval shape of electrodes 38, 39 and the ring antenna 33 are for EM activation of neutral plasma 42 around the spacecraft. The set of electrodes 40 is comprised of one or more narrow flat annular electrodes, completely isolated and not electrically connected to any activating circuit. Their functionality is similar to the electrodes 10 in Fig. 2 - to guide the plasma near the spacecraft body. The activation circuit for the bottom electrodes 39 is a three-phase AC high voltage system based on Tesla coils the secondary of which are of Y type of configuration. The three phases are connected in series with attenuator-phase shifters at the three bottom electrodes 39, while the common point of the Y type is connected to the ring antenna 33. Separate rectifiers for each phase assure the necessary

DC high voltage for the electrodes 39. The top electrode 38 obtains AC+DC high voltage from a separate single-phase Tesla coil [16], also in series with an attenuator. The common point of the Y type 3-phase Tesla coil and the second Tesla coil are connected together forming a "virtual ground" to which the ring antenna 33 is connected. A preferred embodiment of the Y-type and single phase Tesla coils with attenuator-phase shifters is shown in Fig. 10. The dipole antenna 32 for emission of EM wave packets is supplied by a separate circuit, which does not have any common point with the circuits connected to the electrodes 33, 38 and 39. Instead of Tesla coils with air gaps, other type of circuits capable of generating high voltage pulses with short duration could be used. The portholes 41 serve to release a proper gas mixture as preactivated neutral plasma. The portholes are outputs of plasma guide tubes connected to an internal plasma-activating unit or plasma dispensers. The means for creation of neutral plasma are known from the prior art. The preactivated plasma removes the requirement for plasma ionization by the electrodes 33, 38 and 39, so they can operate at lower AC+DC high voltages. When the set of electrodes 38, 39 and 33 are supplied by AC+DC high voltages with proper magnitude and phase, an asymmetric plasma envelope 42 appears around the spacecraft. According to the SARG effect, this will permit creation of a force field in any desired direction. Additionally the gravito-inertial mass of the spacecraft can be reduced even in a motionless position if the activated plasma envelope is symmetrical. In this case the debit of the preactivated plasma must be increased. From this condition the spacecraft could be accelerated sharply from a stationary point or can make a sharp turn during a straight motion by changing the plasma envelope from symmetrical to asymmetrical. This option allows the spacecraft to accelerate with less energy because its mass is reduced at the beginning of acceleration. The gas mixture must contain a working gas of low atomic number and a buffer gas with a larger dielectric strength.

The AC high voltage between the electrodes 33, 38, 39 may not be continuous but in packets. The gravito-inertial effect shows sustainability for a finite time after the cause is removed. The effect has been observed by Kozyrev [17]. He weighed solid objects, then vibrated them and weighed them again. The vibrated solid objects lost a small fraction of their weight and their normal weight was restored exponentially over a few minutes. He found that the restoration time does not depend on the weight of the object but on its density. This is in good agreement with the BSM-SG explanation, that the disturbed self-synchronization needs a finite time for its restoration and that the denser material more strongly affected the CL space inside of the body – a micro effect of General Relativity. The finite time restoration of the CL nodes self-synchronization is convenient for combining the pulse type activation of the surrounding plasma with the creation of the field shield, which also requires repetition of the EM wave packets and the strong superluminal pulse. The other effect found by Kozirev – the dependence of the weight restoration time on the object density – allows selection of the proper material for shielding the crew from the effects of the spacecraft acceleration. When moving in a planetary gravitational field, the crew will feel almost a normal gravitational field, but when the spacecraft is far from any planet, only the local star gravitational field will be felt.

During landing or take-off from a planetary surface the electrode 39 must be at some distance from the ground in order to avoid short circuit or parasite discharge, so retractable legs are needed. The protective field shield must not operate during the landing or take-off, so the dipole antenna 32 must not be supplied in these cases. During maintenance, repair, or staying on the ground, the electrodes 32, 36, 38 and 39 must be grounded for human safety.

The portholes are connected with plasma guides inside of the spacecraft and they carry the preactivated plasma. The means for generating and guiding the preactivated plasma are known from the prior art. The plasma guides suggested by S. Okazaki and M. Kogoma are suitable for this purpose [23]. The electrodes are on the external side of the isolated tube and they could be additionally sealed to prevent the parasite corona discharge.

A spaceship for long range travel with a propulsion system based on the SARG effect must have a different configuration because it will travel a long time in a space with a greatly reduced gravitational field. The side and the oblique view of the preferred embodiment of such a spaceship is shown respectively by Fig. 9.a and Fig. 9.b, where 43 is the overall shape of the spacecraft, 44 and 45 are respectively front and back end thrust actuators, and 46 is a three row set of side thrust actuators an angle of 120 deg between the rows. The preferred motion of this spaceship is along the axis 47, in which case the thrust force is assured by the actuators 44 and 45, creating respectively plasma 48 and 49. If the plasma 48 is stronger than the plasma 49 the motion is from left to right. When moving in a strong gravitational field, the sets of side actuators 46 must also be activated. The two sets of actuators 46 at the angle of

Field Propulsion by Control of Gravity

120 deg are capable of keeping a proper orientation in a gravitational field. In long range travel, gravity from the star system is negligible for the crew, so it is necessary to create it artificially. This can be done by rotation of the spaceship around the axes 47. The necessary rate of rotation can be achieved by a proper phase activation of the 3 sets of side actuators 46, as they are shown in Fig. 9.b. They create side plasma 50. This type of spaceship is not suitable for landing and for this purpose it may carry disc-shape spacecrafts as the one illustrated in Fig. 8. It may not have the maneuverability of the disk-shape spacecraft and must contain a thick shield for protection from harmful cosmic radiation by classical means.

Fig. 10 shows a proffered embodiment of the circuit that supplies the AC+DC high voltage for the three electrodes 39 of the spacecraft shown in Fig. 8 and each triad of the side actuators 46 of the spaceship shown in Fig. 9 a, b. The circuit is comprised of a low voltage AC power supply 51, a pulse transformer 51, a choke 61, a spark gap 53, a group of three primary coils 54 in series with capacitors 55, the said group connected in series with the spark gap 53, a second group of three secondary coils 56 in series with capacitors 57, said secondary group connected from one side to the electrodes 39 and from the other through three groups of parallel capacitors 58 and 59 to a common point 60, said elements 57, 56, 58 and 59 form a Y type three-phase system. Each electrode 39 is connected by a parallel group of resistors 16 and rectifiers 17 to the common point of the Y type system 60. The group of elements 16 and 17 serve to supply the electrodes 39 with a DC high voltage, which is proportional to the AC high voltage. The number of windings of 56 is much larger than the number of windings of 54, according to the Tesla coil considerations in order to obtain the necessary AC high voltage. The elements 59 are high voltage variable capacitors, which serve to adjust the phase difference between the AC high voltages provided to the electrodes 39. They are shunted with capacitors 50 to assure a safety limit in the regulation of the phase, which is accompanied also with a change of the magnitude, causing some asymmetry of the central point 60 in respect to the electrodes 39.

The single-phase Tesla coil for supplying the AC+DC high voltage to the top electrode 38 is comprised of a spark gap 62, a primary coil 63 a secondary coil 64, connected from the bottom side through capacitors 65 and 66 to the middle point 60 of the Y-type three-phase Tesla coil, end connected to the electrode 38 via capacitor 67. The rectifier 17 through the resistor 16 assures the necessary DC high voltage potential for the electrode 38.

Let us not consider at first the effect of the bottom electrodes 39 on the spacecraft shown in Fig. 8. If the regulators 58 of the three phase shifters-attenuators are at equal calibrated positions the phase difference between the 39 electrodes is 120 deg and the voltage magnitudes between them and the point 60 are equal. The plasma surrounding the spacecraft at the bottom is uniform. If the phase difference is unbalanced by the regulators 58 the voltage magnitudes also become different and the plasma envelope on the spacecraft bottom is asymmetric. This will invoke a force field in the direction of the stronger to weaker plasma envelope. When considering the function of the top electrode, it is evident that a force field could be created in any desired direction including the possibility of tilting the spacecraft.

Fig. 11 shows a second preferred embodiment for generation of AC + DC high voltage for plasma activation in a disc-shaped spacecraft. It is comprised of a DC high voltage generator 70 with one pole electrically connected to the bottom side of a secondary of a Tesla coil 68 with a primary 69 and another pole of 70 electrically connected to three bottom electrode modules 71a, 71b and 71 c, said bottom electrode modules 71a, 71b and 71c are mechanically connected respectively to phase controlled motors 72a, 72b and 72c, while the top side of the said secondary Tesla coil is electrically connected to the top electrode 38.

One of the bottom electrode modules 71a, 71b and 71c is shown in Fig. 12. It is comprised of a hollow sphere 73 made of isolation material, inside of which are placed two oval-shaped electrodes 74a and 74 connected by a conductive bar 75, mounted on a non-conductive shaft 76 supported on two bearings 77, the rotation of the shaft 77 is obtained from an external shaft 79 through magnetic coupled flanges 78a and 78b, while the external shaft 79 is mechanically connected to one of the phase controlled motors 72a, 72b or 72c. The sphere 73 protrudes through the spacecraft body having an isolation section 84 around the sphere 73 with an inside metal shield 85 and outside isolated rings 86. The outside section of the protruding sphere 73 is covered by external electrode 82, while on the internal section side there is an electrode 80 connected to the bottom pole of the DC high voltage generator 70 and covered by isolator 81 for preventing a corona discharge. The sphere 70 is made of strong isolation material capable of holding a vacuum, so its internal volume is evacuated through the outlet 83. Technologically, the sphere 70 is made of two hemispheres permitting mounting of the internal members, a feature not shown in the figures.

Fig. 13 shows the timing diagram of the generated DC + AC high voltage for the second preferred embodiment described by Fig. 11 and 12. The lower frequency high voltage 87 is obtainable by capacitive transfer of a high voltage from electrode 80 to electrode 82 due to rotated electrodes 74a and 74b. The higher frequency high voltage 88 superimposed on the waveform 87 is generated by the Tesla coil 68 with a primary coil 69. The primary frequency and the burst rate of the Tesla coil are adjusted to activate the neutral plasma around the spacecraft. The high voltage 87 has a DC component and a low frequency AC component.

The preferred embodiment described by Fig. 11, 12 and 13 has some advantages over the embodiment described in Fig. 10. The first advantage is a more efficient generation of DC high voltage. The second advantage is assurance of a low frequency high voltage component 87, whose phase and frequency is controllable by the phase controlled motors 72a, 72b, 72c. This assures more uniform distribution of the activated neutral plasma around the spacecraft, which is important for the efficiency of the generated force field. The frequency and phase control depend not only on the size of the spacecraft but also on the pressure and constituents of the surrounding atmosphere and the working gas mixture. The third advantage comes from using a classical single phase Tesla coil mounted conveniently in the middle of the disk-shaped spacecraft along its vertical axis, while the high voltage DC generator is placed below it. This shortens the high voltage connections and facilitates their isolation.

The described propulsion method based on the SARG effect is not suitable for application in commercial aircraft flying in Earth atmosphere because of the following reasons:
- the effect creates EM noise and may affect
    communication systems within some range
- staying in proximity to the spacecraft during landing
    or takeoff is dangerous for humans and other living species.

The suggested propulsion method is intended only for spacecraft capable of leaving Earth atmosphere and traveling to other planets or star systems. A spacecraft with such propulsion will greatly outperform rocket missions based on a jet propulsion system.

Chapter 4

I CLAIM:

1. A method and apparatus for spacecraft propulsion with a field shield protection, wherein said method is achievable by an envelope of neutral plasma activated by AC and DC EM fields and simultaneous emission of EM wave packets and strong EM pulses with selected time sequence, said method comprising steps of:

    Releasing of a preactivated plasma of gas mixture composed of working gas and a buffer gas,

    creation of an envelope of electromagnetically activated plasma around the spacecraft by a controllable set of AC and DC electrical fields around the spacecraft from at least 5 electrodes assuring a necessary asymmetry of the plasma envelope, where the said DC electrical fields are kept proportional to the applied AC fields,

    emission of EM wave packet and strong EM pulse with a strong space and time correlation between them,

    said apparatus for a spacecraft propulsion with a field shield protection comprising of:

    a spacecraft body without external sharp edges when in operational mode,

    a set of at least three bottom oval shape electrodes, each one connected simultaneously to a Y type three-phase AC high voltage system with a proper frequency and means for phase and amplitude control,

    at least one top electrode connected simultaneously to a single phase AC high voltage system and to a DC high voltage,

    said single phase high voltage system having a common virtual ground with the central point of the said three-phase AC high voltage system,

    separate DC high voltage circuits for each of the bottom and top oval shape electrodes, obtained by rectifying a fraction of the AC high voltages that supply those electrodes,

    one ring electrode comprised of one or more sections separated by one or more gaps whereas at lease one of these sections is connected to the virtual ground point of the said three-phase AC high voltage system,

    one circular dipole antenna connected to a generator of EM wave packets with a circumference length of the dipole

101

antenna equal to one or multiple wavelengths of the said EM wave packet,

a set of portholes on the spacecraft body for releasing of preactivated plasma from a gas mixture, where said gas mixture is comprised of a working gas with a low atomic number and a buffer gas with a higher dielectric strength.

2. The invention of claim 1, wherein the propulsion force field is a result of a gravito-inertial effect invoked by applying an asymmetrically activated neutral plasma around the spacecraft.

3. The invention of claim 1, wherein said protective field shield is a result of mutual interactions between the emitted EM wave packet and the strong EM pulse possessing a superluminal behavior with proper time correlation between them and a proper period, so an effect of energy dissipation occurs on the molecular species or dust particles in the boundary of finite thickness serving as a protective shield.

4. The invention of claim 1, wherein said Y type three-phase AC high voltage system is part of a poly-phase Tesla coil system, in which the secondary windings are connected as Wye type, while the primary windings are connected in series with capacitors and one spark gap or other type of circuit interrupter.

5. The invention of claim 3, wherein said strong electromagnetic pulse possessing superluminal behavior in a near field of a few wavelengths is experimentally confirmed.

Chapter 4

REFERENCES: (for this patent application)

[1] Albert Einstein, *Sidelights on Relativity*, Methuen & Co. Ltd, 36 Essex Street, W. C., London, 1922 (Republished later)
[2] Michelson, A. A., Morley, E. W. (1887). "On the Relative Motion of the Earth and the Luminiferous Ether". *American Journal of Science* **34** (203): 333–345.
[3] S. Marinov, Measurement of the Laboratory's Absolute velocity, General Relativity and Gravitation, **12**, 57-66, (1980)
[4] S. Marinov, The interrupted "rotating disc" experiment, J. Phys A: Math. Gen, **16**, 1885-1888, (1983)
[5] S. Marinov, New measurement of the Earth's Absolute Velocity with the help of the "Couplet Shutters" Experiment, Progress in Physics, v. 1, , 31-37, (2007) (description of one Marinov's experiment done in 1984, but this article is submitted by Erwin Scheerberger 10 years after Marinov's death)
[6] S. Sarg, New approach for building of unified theory, http://lanl.arxiv.org/abs/physics/0205052 (May 2002)
[7]. Stoyan Sarg, *Basic Structures of Matter – Supergravitation Unified Theory*, Trafford books, Canada ISBN 1412083877, (2006)
[8]. S. Sarg, A Physical Model of the Electron according to the Basic Structures of Matter Hypothesis, Physics Essays, International Journal Dedicated to Fundamental Questions in Physics, vol. 16 No. 2, 180-195, (2003); http://www.physicsessays.com
[9] S. Sarg, Basic Structures of Matter – Supergravitation Unified Theory based on an alternative concept of the physical vacuum, Proceedings of IX International Scientific Conference " Space, Time, Gravitation, 7-11 Aug, 2006, Saint-Petersburg, Russia
[10] Books review, in "**Physics in Canada**,", v. 62, No. 4, July/Aug, 2006, published by the Canadian Association of Physicists)
[11] S. Sarg, Gravito-inertial Propulsion Effect Predicted by the BSM - Supergravitation Unified Theory, 27th Annual Meeting of the Society of Scientific Exploration,
June 25-28, 2008, Millenium Harvest House, Boulder, CO
[12] Massines et al. Experimental and theoretical study of a glow discharge at atmospheric pressure controlled by dielectric barrier, J. Appl. Phys. 83, 2950 (1998)

[13] K. J. van Vlaenderen and A. Waser, "Electrodynamics with the scalar field, www.aw-verlag.ch/EssaysE.htm also with slight adaptations: van Vlaenderen Koen and A. Waser, "generalisation of classical electrodynamics to admit a scalar field and longitudinal waves", Hadronic Journal **24**, 609-628 (2001)

[14] K. P. Butusov, Longitudinal Waves in Vacuum: Creation and Research, New Energy Technologies, Sep-Oct 2001, pp. 46-47.

[15] Lord Kelvin, On the generation of longitudinal waves in Ether, Proceedings of the Royal Society of London, v. 59, pp. 270-273, (1895-1896)

[16] Nicola Tesla, Colorado Spring Notes 1899-1900, NOLIT, Beograd, Yugoslavia, 1978.

[17] Н. А. Козырев, Избранные труды, Ленинградский Государственный Унивирситет, 1991

[18] Козырев Н. А, Насонов В. В, Новый метод определения тригонометрических параллаксов на основе измерения разности между истинным и видимым положением звезды // Астрометрия и небесная механика М.; Л., 1978. С.168-179. (Проблемы исследования Вселенной.Вып.7).

[19] Лаврентьев М.М., Еганова И.А., Луцет М.К., Фоминых С.Ф. О дистанционном воздействии звезд на резистор // Докл. АН СССР. 1996. Т.314, № 2. С.352-355.Лаврентьев М.М., Еганова И.А., Луцет М.К., Фоминых С.Ф. О регистрации реакции вещества на внешний необратимый процесс // Докл. АН СССР. 1991. Т.317, № 3. С.635-639.

[20] Hodowanec Eclips of 10-3-1986

[21] D, Mugnai, A. Ranfagni, and R. Ruggeri, Observation of Superluminal Behaviours in Wave Propagation, Phys. Rev. Lett., v. 84, No 21, 4830-4833, (2000).

[22] T. W. Barrett, Tesla's nonlinear oscillator-shuttle-circuit (OSC) theory, Annales de la Fondation Louis de Broglie, V. 16, No 1, 23-41, (1991)

[23] S. Okazaki and M. Kogoma, J. Photopolym. Sci. Technol., Vol. 6, No 3, 1993.

Search keywords in google and youtube: SARG effect, SARG Antigravity.

Chapter 4

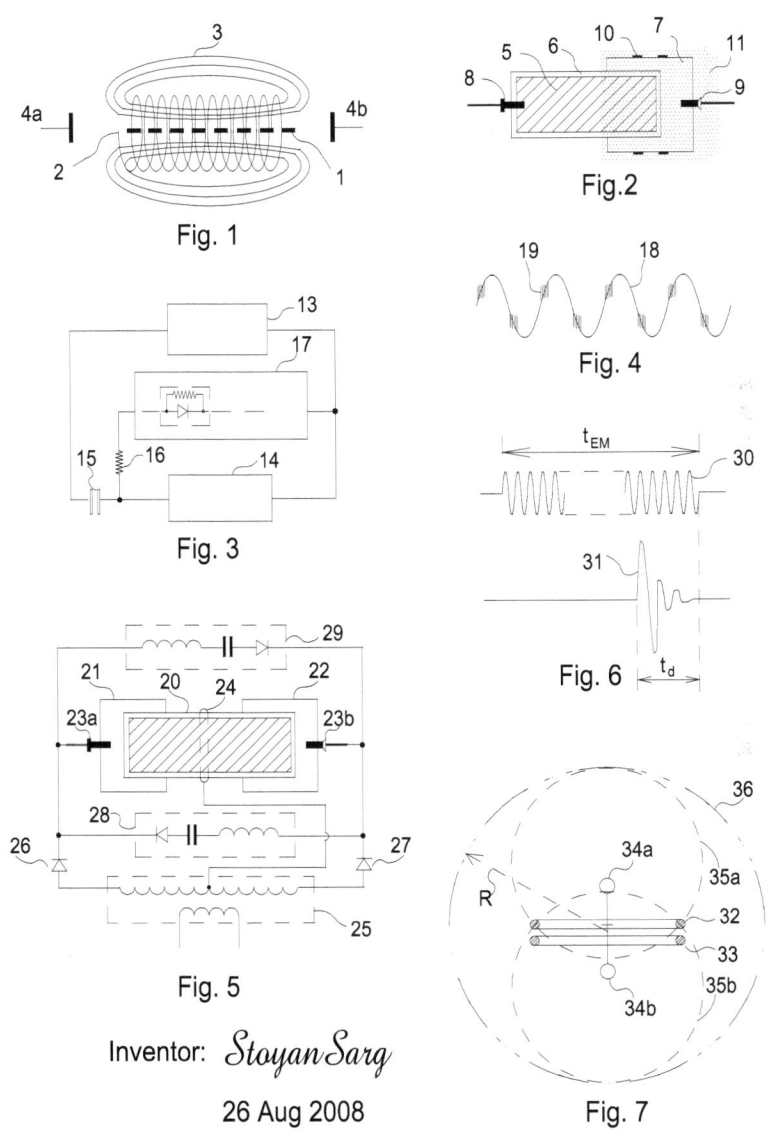

Inventor: *Stoyan Sarg*
26 Aug 2008

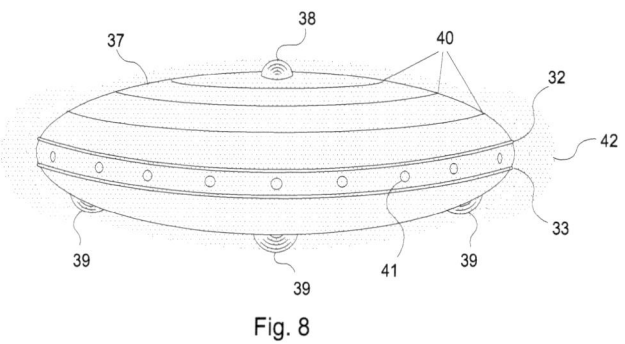

Fig. 8

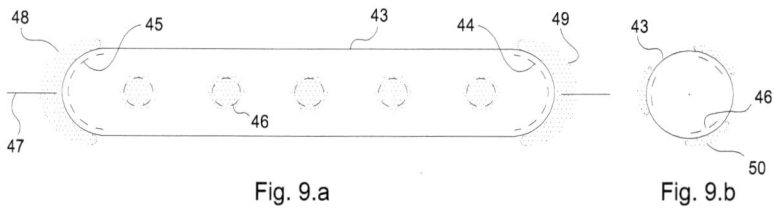

Fig. 9.a Fig. 9.b

Inventor: *Stoyan Sarg*
26 Aug 2008

Chapter 4

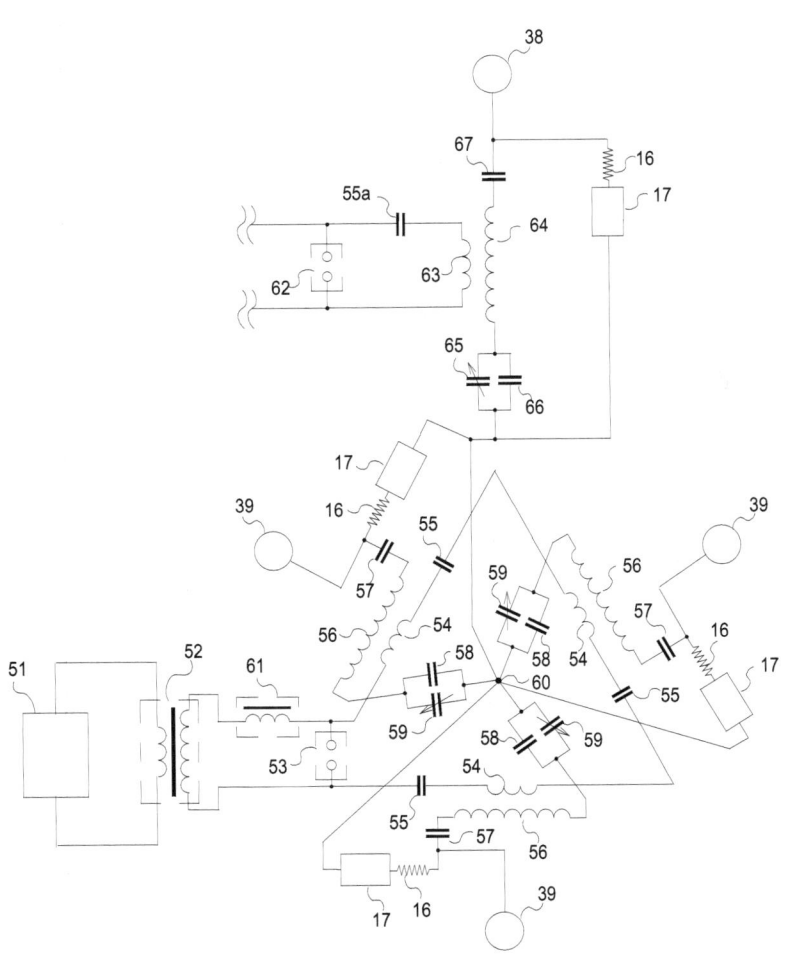

Fig. 10

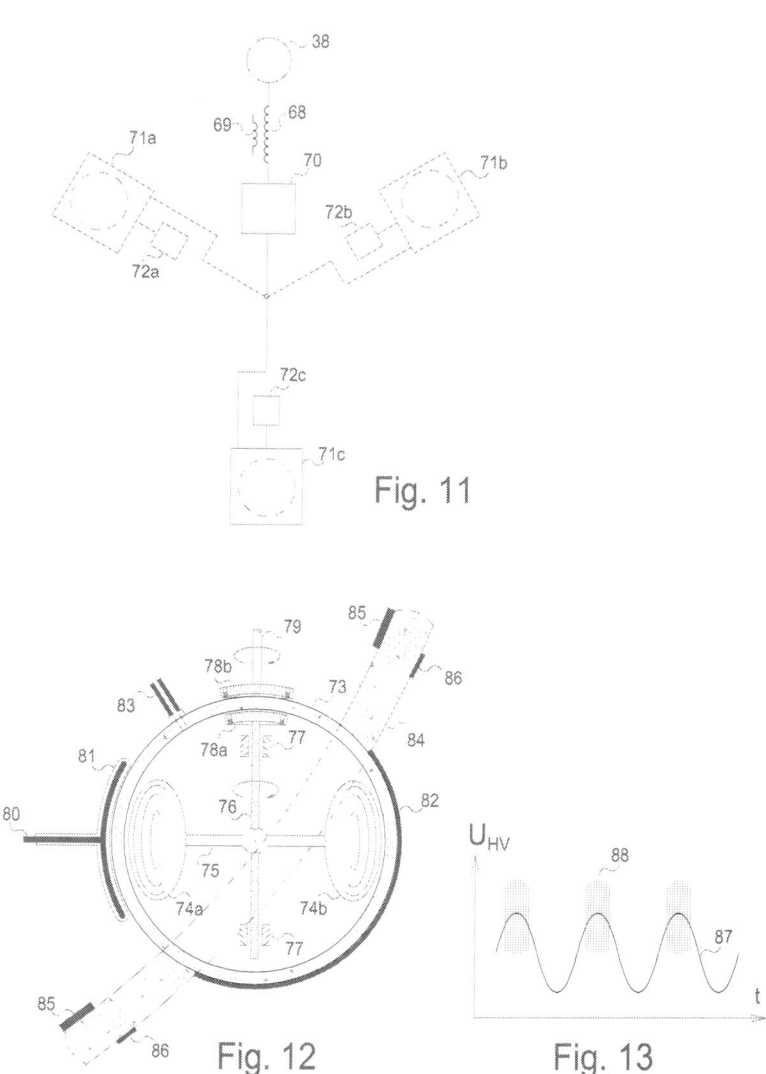

Fig. 11

Fig. 12

Fig. 13

Inventor: Stoyan Sarg

CHAPTER 5. Field shield protection and Supercommunication: Hypotheses.

5.1. Field shield protection: Hypothesis.

A spacecraft with a propulsion system based on the SARG effect needs a field shield for:
- Protection from micrometeorites in deep space when traveling with a large velocity
- Protection against a threat from an enemy

The concept of a field shield protection is envisioned from a BSM-SG point of view and the analysis of some published experiments and observations. The author, however, has not tested this concept since it requires quite special environment and equipment.

The predicted technical realization of field shield protection was described in Chapter 4. Here we will highlight only some predicted features and its protection capability.

Field shield protection should be achieved by a proper combination of EM and Longitudinal Waves (LW) emitted in packets with the necessary timing characteristics (see Chapter 4, Figures 6 & 7). It is based on the BSM-SG prediction that, while the isotropic LW attenuate quickly with distance, they travel with a superluminal velocity. This is in agreement with the astrophysical observations provided by N. A. Kozyrev [29] and experiments detecting superluminal wave propagation in proximity to a LW emitter. [43]. The spacecraft will be able to generate a field shield having a spherical shape by emitting EM and LW packets with proper timing between them as discussed in Chapter 4. To explain the concept we must refer to several unveiled features of CL space, related to propagation of EM waves:
- Ferromagnetic hypothesis discussed in BSM-SG Chapter 8, §8.16.
- Dynamical Property of the Cosmic Lattice. MQ and SQ SPM vector. (BSM-SG, Chapter 2, §2.9)
- Quantum wave configuration (BSM-SG Chapter 2, §2.10.4.2)

According to the Quantum wave configuration of the photon, it preserves its quantum energy during propagation due to the boundary conditions of its wave train structure. The boundary conditions result from two types of modulations of the SPM (Spatial

Precession Momentum) vector that describes the photon wave train. The two types of SPM modulations are defined by the hodograph of the SPM vector as a MQ (Magnetic Quasisphere) modulation and an EQ (Electrical Quasisphere) modulation. EQs and MQs have a helical arrangement in the wave train and they define respectively the time and space propagation features of the H and E vector of the EM field. In isotropic radio waves there are no boundary conditions, but the MQs and EQs of the SPM vector are distributed in the expanding wave train volume.

An important feature of the MQs in the photon and EM waves is that they are phase synchronized at the speed of light. In other words, the phase difference of the MQs and EQs involved in the photon or EM wave, propagated at the speed of light, is constant. Then we may raise a question: what will happen if the MQs involved in an EM wave are synchronized at a speed faster than the speed of light? We have already mentioned that the Longitudinal Waves (LW) emitted from a single emitter are able to propagate with a superluminal velocity [43] if their energy is above some threshold level. If such an emitter is not in resonance with another emitter or receiver, the LWs will attenuate quickly with distance. However, if the intensity of the LWs is above some critical threshold level, they could affect EM waves if both types of waves are emitted from the same source. More specifically, the LWs will affect the synchronization between the MQ nodes. Instead of a phase synchronized at the speed of light, they will be synchronized at a greater speed. The intrinsic frequency of the SG forces is several orders higher than the resonance frequency of the CL node, but their propagation is attenuated by the proper resonance frequency of the CL node. If the phase difference between the EQs and MQs of the EM wave is not constant but changed in the proper direction, a strong attraction may occur between CL nodes that will affect the distribution of the EQs and MQs. The predicted result is a space-time concentration of MQs, which means a reduced length of bound magnetic lines and, consequently, a shorter duration of the wave train. If the decrease in duration is towards the center of the EM pulse duration, then it will lose the momentum for propagation at the speed of light. This will lead to a local change in the CL space parameters ε_0 and μ_0. This alters the optical properties of the physical vacuum, thus forming a boundary on the direction of wave propagation. If the EM waves and the LWs are emitted in bursts with an appropriate time delay as discussed in Chapter 4, the consecutive bursts will reinforce the formation of such a boundary. For a single isotropic emitter of EM and Longitudinal waves, the boundary will have an oval shape with a sharper gradient on the external side.

How might the protective field shield work? In §7.7.1 of BSM-SG, the conditions for a stable quantum orbit are discussed. It was found that, for a stable quantum orbit with a lifetime corresponding to spontaneous emission, the magnetic environment must be stable. So if the magnetic environment is changed, the electron will drop to a lower orbit. When the permeability of the physical vacuum is severely disturbed by the creation of fragmented zones, a body passing through such zones will undergo strong internal stresses as many bond connections between the atoms or molecules are broken. This effect will become stronger as the velocity of the object increases. If the object is a micrometeorite, it could easily disintegrate. For a larger object, a stronger protective shield would obviously be necessary. However, the protective field shield should not be operational all the time. It must be turned off during landing and take-off, as well as while landed if a crew is outside. The field may be very harmful to humans or other living biological entities. In some UFO reports, humans in close contact with the object suffered serious or fatal health consequences. One such case (The Falcon lake incident) is described in the book "The Canadian UFO report", authored by C. Rutkowski & G. Dittman [53].

A number of UFO observations lead to the conclusion that they have a protective field shield but it is rarely activated. When activated, it should be observed in twilight or in nighttime. A few such observations have been reported. It must be mentioned that in the case of three coherent emitters of EM and LW, flashing should be observed at some distance from the UFO but centered on the central axis. An interesting feature observed in a number of cases is a light beam with an abrupt termination. A picture of a UFO emitting three light beams with such terminations is shown in Fig. 5.1. A light beam with such an abrupt termination is a mystery, although BSM-SG theory offers a hypothetical explanation. The technical realization of such a phenomenon could be similar to the method of proper combination between EM waves and LW discussed earlier. In this case, however, the EM waves are in the visible or infrared range and the LW must be focused at infinity.

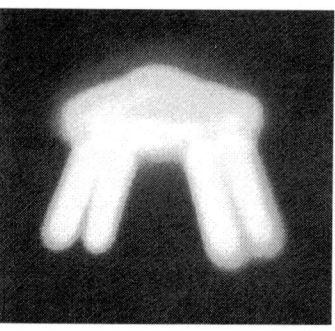

Fig. 5.1. An aerial craft projecting beams of light, photographed by an anonymous doctor on 23 March 1973, near Tavernes, Var, during a wave of sightings in France. Adapted from Timothy Good book Alien Base [56]

Focusing of LWs is a completely new problem that will not be discussed here. It must be said only that it requires different techniques for the radio and optical regions. In the latter case ultrashort laser pulses could exhibit LW properties.

Here we must emphasize again that a spacecraft with the described new propulsion mechanism cannot replace airplanes. It is only for interplanetary or interstellar travel.

5.2 Unconventional method of communication by Longitudinal waves: Hypothesis.

For a spacecraft with a propulsion system based on the SARG effect, ordinary EM waves are not a suitable medium for communication between the crew and the outside world. This is particularly valid for spacecraft with the shape of a "flying saucer" since the activated plasma may wrap the whole spacecraft.

Considerations why the conventional EM waves cannot be used are as follows:
(1) The plasma surrounding the spacecraft disturbs the propagation of EM waves.
(2) The antenna for EM communication must be inside the spacecraft but it behaves as an ideal Faraday cage.
(3) The propulsion systems of some type of spacecraft might emit a large amount of longitudinal waves that for ordinary EM receivers will cause broadband EM noise (like the noise from lightning).

(4) The disturbance of the surrounding space by affecting the self-synchronization between CL nodes changes not only the inertial properties of the spacecraft but also the average proper frequency of the affected CL nodes. Since the self-synchronization disturbance from the spacecraft is dynamical, the time etalon of the surrounding zone will fluctuate dynamically. A Doppler fluctuation from the spacecraft motion will also be added. This means that ordinary EM waves passing through this zone will fluctuate both in frequency and amplitude. For the external receiver, such EM waves will have random-like frequency variations and amplitude fluctuations. The same is true for the EM waves received by the spacecraft crew. Evidently, such EM communication will be highly unreliable.

More reliable communication would be achieved by using transmitters and receivers of longitudinal waves. This would require special converters from EM to LW and from LW to EM waves.

The LWs would need to traverse the Faraday cage of the spacecraft, the surrounding plasma and the field shield protection. They would not suffer from the frequency disturbance in the zone of surrounding plasma and the protective field because they are based on compression of the CL space (physical vacuum).

The properties of some types of LWs were discussed in section §1.5 of this book. Propagation of the Isotropic LWs is different from propagation of EM waves. Amongst their advantages over the EM waves is the ability to propagate with a superluminal velocity and pass through a Faraday cage. The disadvantage, however, is the fast attenuation with distance. In order to be used for communication, some new category of this type of wave must be created: like the photon, they must preserve their energy. Further, the LWs should be in the radio-frequency range for easier conversion to and from EM waves.

The photon wavetrain as a quantum wave is discussed in §2.10.4, Chapter 2 of BSM-SG. It possesses straight-line propagation and boundary conditions that prevent dissipation of the transported energy momentum. The EM wave in the radio frequency and microwave range does not have such boundary conditions even in a focused beam, so its energy more or less dissipates with distance. The boundary conditions of the photon are created by the boundary CL nodes, the behavior of which is described by the MQSPM vector synchronized by phase.

Field Propulsion by Control of Gravity

Consequently, we would like to create a unidirectional LW propagated in a straight line with a wave train having a finite radial cross-section surrounded on the outside by MQSPM CL nodes. This could be achieved if the following considerations are fulfilled:
  (a) the magnetic field included inside of the unidirectional LW must have a boundary limit or be compensated at some radial distance from the central axis.
  (b) the internal section (core) of the device (coil or electrode) creating the LWs must be of denser material.
  (c) from (a) and (b), it follows that the internal permeability of the device that creates the unidirectional LWs must be different from the external permeability.
  (d) Creation of LWs requires strong nonlinear interactions. This means that the generating power must exceed some threshold.

The consideration (d) is similar to obtaining shorter wavelength in lasers (2nd or 3rd harmonic) by illuminating a crystal. A signal intensity threshold must be exceeded in order to obtain a shorter wavelength.

5.3. Wilbert Smith and his Tensor Coil (Smith Coil).

Following the above considerations, one possible device for creation of unidirectional LWs could be the Smith coil. However, there is not yet enough experimental research on it.

Wilbert Smith (1910-1962) was born in Lethbridge, Alberta Canada. He obtained his B.Ss and M.Ss degrees in Electrical Engineering, from the University of British Columbia, Canada in 1933 and 1934. He worked for four years as a Chief Engineer with Radio Station CJOR, Vancouver, B.C. and in 1939 he joint the Department of Transportation of Canada. Mr. Smith advanced in broadcasting with growing responsibility and scope over the years with special tasks from the Canadian government. In 1959 he was authorized to lead Project Magnet sponsored by the Department of Transportation. Laboratory facilities were created to study the phenomenon known as UFO – or Unidentified Flight Objects. Four years later the Department of Transportation dropped the project. However, Mr. Smith continued his interest and worked in the UFO field privately. One of Mr. Smith's known inventions (not patented) is a "tensor coil" as he called it. This device is known today as a "Smith coil". The following is extracted from a talk given by Mr.

Chapter 5

Smith at the Vancouver Area UFO Club, March 14, 1961. It was entitled "What we are doing in Ottawa." (search Internet by these keyword).

Now one other thing that I would like to mention, as far as I know our group in Ottawa is the only group that has actually taken the info, from the boys topside and translated it into hardware that works. Much info has been given to us through other channels, but people just talk about it. They don't do anything about it and I think that's deplorable. I think when they give info, the least we can do is to show our good faith by trying to convert the info into hardware. We have built two items of hardware on their instructions that I'm rather proud of. One is a coil. It has a ferrite core and a trick winding on it. To look at it, it looks rather like a rather oddly wound inductor. When measured on a radio-frequency bridge it shows some very peculiar properties. There are certain frequencies at which it is impossible to balance the r.f. bridge, and that is a direct contradiction to what any electrical engineer will tell you should happen with a coil of wire wound on a ferrite core. Now if we take this coil and we excite it with radio frequency energy at or near these critical frequencies, we find that energy goes into the coil and nothing comes out. It's just disappearing. As a matter of fact we had one coil about an inch in diameter and eight inches long and we poured a kilowatt into that coil for two hours from a kilowatt communications-type transmitter. The coil was in a two-inch brass tube with a plate welded on one end and a transmission line fitting on the other. We could find no radiation around the outside of that tube at all. In other words the energy went in; now came out. The info, which we got from the boys topside was that we were making tenser energy, which is a sixth-dimensional radio wave, and is the type of energy they use extensively for radio communications, transmission of power and for pushing and pulling. In fact they use it for just about everything that we could think of. We were not able to control this energy; we could just make it. We are hoping that later on we will be able to learn how to do it, but at the present time we are not just smart enough.

Field Propulsion by Control of Gravity

The following is part of a letter from Mr. Smith to Mr. Middelton, January 2, 1955.

We are told about a system which uses a radio transmitter as an energy source but a special antenna converter, which radiates doughnut shaped waves, which are not time fluctuations. We built a couple and find they have the most extraordinary properties.............Following are the construction instructions.

One ferrite core, material with a highest possible permeability and dielectric constant, about 8 inches to a foot long, and about 1 inch in diameter. About 20 feet of plastic insulated #14 (gage) electric house wire. Starting at the center of the wire at one end of the core wind the wire as closely as possible, with the first turn under and then over, so that the winding will be exactly symmetrical. It will start at one end of the core and finish at the other end and will resemble a solenoid with a bifilar winding. It is important that winding be exactly symmetrical.

When connected to a transmitter, treat it as any normal antenna for loading and tuning. There will be a few points of magnetic domain resonance witch will be lossy but anywhere else the device will generate the required waves. It will not matter whether or not the antenna converter is shielded as the doughnut waves go through anything...

One of our groups here George La Fleur has had quite a bit of success in reaching many amateurs, and I shall ask him to try and contact you on the 14 meter band.

In another letter to David Middelton, January 11, 1959, Mr. Smith writes about his coil: *It is a converter of energy from ordinary radio frequency to tensor.*

Another clear description of the Smith coil is published by Gaston Burridge, an editor in Fate magazine, who spent great efforts to collect information about this coil from the collaborators of Mr. Smith ten years after his death. The following is an extract from the G. Burridge article called " The Smith Coil" published in Fate magazine.

Chapter 5

The coil winding is comprised of insulating wire – either double cotton-covered or enameled. Generally, the size of the wire has been given as #18 gage copper – although aluminum, iron and even silver have also been used. The coil is caduceus wound with two wires – the wires opposing each other around the core, and these two wires opposing each other on the two opposite sides of the core's diameter with each complete turn. Hence, the coil, when completed, will have two rows of bumps on it. These bumps will be opposite one another and on each side of the coil, formed where the two wires cross. Great care should be taken in making this winding exact, so these "crossovers" (bumps) remain in a straight line along opposite sides of the core."

In another letter to Mr. Williams, May 9, 1955, Mr. Smith writes:

We used one inch Ferrite tubes with a half inch hole, about 8 inches long and plastic insulated wire. The theory of operation is that the RF current flowing in each turn generates loops of magnetic flux within the ferrite, and this in turn generates loops of electric flux and the whole is threaded on loops of tempic field. As the radio currents progress along the ferrite rod the little doughnuts are pushed off the end and sent on their way. Reception takes place whenever the doughnuts encounter any material through which they can progress with the same net phase conditions as if they passed outside of it. In other words it is a three field resonance condition.

Wilbert Smith published a theory called "The new science". It is available on-line at http://www.rexresearch.com/smith/newsci.htm

He speaks about space fabrics with features amazingly similar to the CL space according to BSM-SG. The BSM-SG theory was first published in 2001 and the author was not aware of W. Smith's theory at that time since it appeared later on the Internet. Smith's letters held in boxes at NRC Canada were released in 2006. The theory of Wilbert Smith and the BSM-SG theory are based on different approaches, but they lead to quite similar conclusions about the space fabric (physical vacuum).

From the preserved correspondence of Mr. Smith, it is evident that he was not officially supported by academic physicists, and they did not acknowledged his unique research. Nevertheless he communicated privately with a number of university professors.

The following extract is from a letter by Mr. Smith to Prof. Cullwick, May 9, 1958.

I am informed by the people from "elsewhere" that our universe is actually twelve dimensioned, built of four fabrics of three dimensions each. The basic reality is spin, which is a fundamental concept and from which all is obtained. Our space fabric comes from the structure of spin.

The above statement about "four fabrics by three dimensions each" matches the configuration of the CL structure according to BSM-SG:

CL node is comprised of 4 twisted prisms = *four fabrics*
EM field is propagated by the xyz axes of the CL node = *three dimensions*

Further from the same letter:

The people from elsewhere are quite capable of altering the time field in their vicinity to their own ends, much the same as electrical and magnetic fields may be manipulated; and therefore, having three variables so the saucers can do things which seam to contradict our physics, because we hold as constants some of the things which they vary.

According to the above-mentioned extracts, the clock rate inside the spacecraft and in the surrounding zone could differ from the outside world. This agrees with our consideration (4) in section 5.2.

At the end of the same letter Mr. Smith says:

Of course, at the best we are still in the kindergarten in this matters, and it may be many long years but at least we now have the clues.

In another letter to Mr. Burridge, Wilbert Smith writes:

Unfortunately, the models of atoms, etc. which we conjure up are inadequate and misleading and conventional science which is based on them, and using restricted theories, naturally can't take us very far.

The above statement is in agreement with the finding of the BSM-SG theory about the atomic models. The real physical models (as those suggested by BSM-SG) must first be described by their material structure and spatial arrangement of the elementary particles in the atomic nuclei. These structures are the hardware and they define the energy levels calculated by Quantum Mechanical models and possible wavefunctions.

With respect to the correct design of a spacecraft with the described new propulsion system, understanding of the material structure of atomic nuclei is important for two reasons: ensuring the necessary design strength and resistance of the spacecraft to LWs, and designing a protective field shield.

Another possible method for detection of LWs used for communication is the gravity detector of Gregory Hodowanec discussed in Chapter 1, §1.7. This detector, however, is for isotropic LWs and may not detect the specific LWs emitted by the Smith coil.

5.4. Some tests of the Smith Coil.

The information about the Smith Coil comes mainly from his letters and reports. The author of this book did not find details about electrical parameters of the coil or of test results. Although the coil looks simple, its operation may need careful selection of the design parameters (shape, magnetic permeability of the ferrite core and number of windings) on one hand, and a proper electrical drive on the other. A successful test of the coil depends on this knowledge. Currently not enough research on this issue has been done. Some experiments are provided by J. L. Naudin (France) but only with a single unshielded Smith Coil. Some researchers have confirmed Wilbert Smith's statement that the coil has unusual behavior in that it resonates over a large frequency range, something impossible for a classical resonance LC circuit. The author of this book also observed this feature when testing a Smith Coil without a ferrite core.

With respect to solving the problem of communication between the spacecraft crew and the external world, it is necessary to focus the research on a system of two Smith Coils – one for transmitting and one for receiving. The goal is to obtain a transmission of non-EM waves (Longitudinal Wave) between two Smith coils. They must be in metal enclosures to eliminate the possible transmission of ordinary EM waves that may contaminate the results. For this reason, the author prepared a pair of Smith Coils with removable ferrite cores and enclosed in iron cylinders. The disassembled view of one of the coils is shown in Fig. 5.1.

Field Propulsion by Control of Gravity

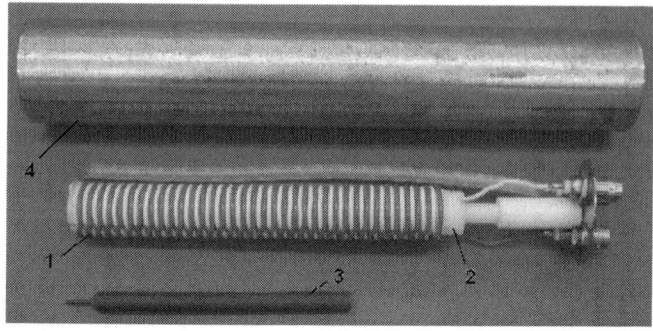

Fig. 5.1. Disassembled Smith Coil; 1 – wiring, 2 – insulator body with a hole, 3 – ferrite core, 4 - steel cylindrical enclosure.

The coil has a length of 14.5 cm and a diameter of 1.8 cm. The total number of pair windings is 36. The copper wire diameter is 0.5 mm and, with insulation, it is 1.5 mm. The central ferrite rod has a diameter of 9.5 mm and a length of 10.3 cm. It is from surplus with unknown specifications and is probably from a ferrite antenna in an old AM receiver. The measured inductance between the b and c terminals is: 5 uH for the coil without the ferrite core; 23 uH for the coil with the ferrite core. The impedance between terminals *a* and *b* is only an active resistance – no inductance is measured.

Fig. 5.2 illustrates the Smith Coil system that was tested. The receiver coil has two terminals for connection to a sensitive differential amplifier. The transmitter coil has three terminals permitting different connections. Both coils and their windings are identical.

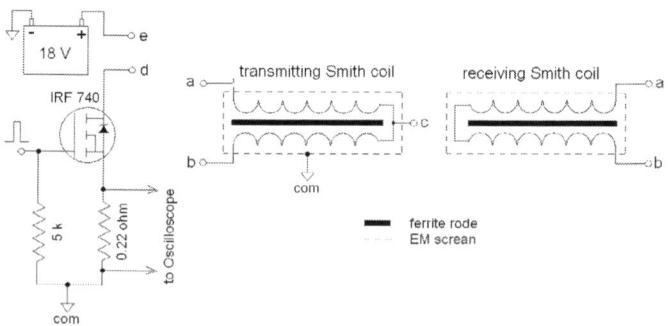

Fig 5.2 Smith Coil test circuit. Transmitting coil connection: First test option: **e** connected to **a** and **d** connected to **b**. Second test option: **a** and **b** connected to **e** and **d** connected to **b**.

Chapter 5

From the point of view of BSM-SG theory, Longitudinal waves (LWs) should result from a sudden change in the electrical field. In the case of the coil, this means that a sequence of strong but short current pulses must be applied. For this reason, we should not expect that the coil would emit LWs if a sinusoidal voltage were applied. We must investigate a specific transient mode. It is known that the coil inductance prevents a fast rise of the current, so we investigated the case of FORCING OF STRONG CURRENT PEAK (current surge) through the coil.

One may expect that the Smith Coil is a kind of bifilar type and should not have impedance for a current pulse. However, a real test of the Smith Coil with a ferrite core showed one interesting feature. Despite the fact that the measurement between terminals a and b showed a zero inductance, when applying a strong short pulse the current appeared limited as in the presence of an inductance.

For a short current pulse, the Smith Coil exhibits an active impedance (active resistance). This is an unusual effect that should be investigated. While a sequence of current pulses must be conveyed to the transmitting Smith Coil, the receiving Smith Coil is expected to deliver a sinusoidal signal.

During the tests provided by the author using the arrangement in Fig. 5.2 with the shielded pair of Smith Coils, there was no detection of LWs passing through the iron shields. Some of the following issues might account for the unsuccessful result:

(a) The ferrite core with unknown magnetic properties may not be suitable.
(b) The power supply was only 18 VDC from a lead acid battery. The experiment may require a much higher voltage.
(c) The coil was made with a uniform winding step defined by the insulation diameter of the wire. The transmitting coil may need to have a linearly increasing step in order to break the axial symmetry.

For researchers who plan to investigate this coil, we offer the following recommendations:

(d) A common ground between the shields and the electrical circuits of the transmitting and receiving coils should be avoided, since some signal may pass through it.
(e) The terminals a and b of the receiving coil should be connected to a differential amplifier (or a scope with differential input)
(f) It is reasonable to try a test of a Smith Coil with a linearly increasing or decreasing winding step.

(g) The receiving Smith Coil has a broadband resonance range but the ferrite has a limited working frequency range that should match the repetition rate of the current pulses conveyed to the transmitting coil.

(h) The experimenter should stay away from the axis of the operating Smith Coil. The biological side effects from the LWs are still not investigated.

Conclusions:

(1). The possibility for Field shield protection is envisioned by the theoretical predictions of BSM-SG theory and the analysis of one of the Hutchison effects, more specifically the structural change of some samples.

(2). The predicted supercommunication is based on the BSM-SG theory and some experimental evidence for superluminal propagation of Longitudinal Waves.

(3). Currently the author is not aware of ongoing research on methods of field protection and supercommunication.

(4). Supercommunication will require a transmitter and receiver based on principles that differ from those for EM communication. The system should also include converters from EM to LW waves and vice versa, which could also be a new type of device.

(5). There is not enough research on the possibility of using the Smith Coil for supercommunication, and especially for passing the signal through the shielded metal body of the spacecraft.

(6). The field protection could be dangerous for living organisms in some particular range. It must be used only in deep space or at altitudes above normal commercial flights.

(7). When using devices for supercommunication based on LWs, proper shielding will be needed for the crew*.

* Note: Based on some experiments, the author concluded that protection from LW waves requires a different type of shielding. For EM waves, the shielding metal screen must be grounded. For LW waves, this method does not work. The LW waves could be more effectively shielded if enclosed in a chamber comprised of a few metal chambers one inside the other. They must be separated by non-conductive isolator material and not electrically connected to each other or to ground. The metal chambers must be oval or, if rectangular, must not have sharp edges.

Chapter 5

SUMMARY AND CONCLUSIONS

The author spent a few years doing research on Field Propulsion based on the gravito-inertial effect predicted by the BSM-SG theory and its realization by the Heterodyne Resonance Method. Initially, the prior art of plasma experiments were reviewed and analyzed using the BSM-SG physical models. The author focused particularly on experiments in the fields of Electrogravity, electrodynamics, plasmodynamics, plasma trust actuators, paraelectric actuators, Hall effect plasma thrusters and so on. Amongst the advanced researchers in USA are the research groups of J. R. Roth at the University of Tennessee and S. Roy at the University of Florida. Despite the extensive research performed by these and other groups of researchers, no one other than the author of this book conceived the idea that the observed effect has a signature of mass change. The mass change affects not only the objects surrounded by the plasma but also the surrounding gas molecules. The observed gas flow and the reduced turbulence surrounding the plasma actuators are in fact signatures of a mass change. The observed thrust appears enigmatic since it contradicts Newton's classic laws of motion when considering the whole system analysis. For this reason the researchers usually avoid such analysis, mentioning only that the underlying physics is not well understood. Therefore they derive only empirical physical models based on analysis of the experiment. Such models, however, are too dependent on the design parameters, so advances in this field have been quite slow. Without a fundamental physical explanation, optimization of the effect was not possible. Another disadvantage of the prior research is that the researchers did not envision that such a propulsion effect could be used in deep space. They considered it could work only in an atmospheric environment. This is a wrong conclusion because of the following two reasons. First, the new propulsion mechanism is not based on inertial interaction with the surrounding air, so an atmospheric environment is not a necessary condition. Second, the new propulsion effect has hidden side effects that could be harmful to the surrounding environment within a limited range. Consequently, the new propulsion mechanism is suitable for space travel but not for commercial use. That is, it could not replace current airplane technologies.

Apart from analyzing the prior art experiments from the point of view of BSM-SG theory, the author built a number of Lab experiments on glow discharge in a partial vacuum and at normal air pressure. This led him to the discovery of a gravito-inertial effect that he called a Stimulated Anomalous Reaction to Gravity (SARG). He recognized that the potential application of this effect is a new kind of propulsion method suitable for deep space travel.

For practical applications of the SARG effect as a new kind of propulsion mechanism, the author provides the following conclusions and recommendations:

(1) The application of the SARG effect for a spacecraft propulsion system requires a specific design for the shape and geometry of the spacecraft with means for creation of a partial to full plasma envelope around it. For maneuvering such spacecraft, active control of the plasma envelope is needed.

(2) Small spacecrafts may be fully enveloped by plasma. They are more convenient for atmospheric environments, while also being able to travel outside of the planetary atmosphere.

(3) Large spacecrafts with an oval cylindrical shape are suitable for distant space travel between planets and stars but are not suitable for frequent landing. They may not be fully covered by plasma but may contain a number of portholes with activation electrodes along their length. The electrodes activating the plasma may work in sequence due to the finite time restoration of the self-synchronization of the CL space (physical vacuum).

(4) The SARG effect affects not only the mass of the actuator or spacecraft but also the mass of the surrounding air molecules. For this reason a reduced turbulence is observed in some experiments [49] and some observations of UFO objects analyzed by the NASA researchers Paul Hill [30] and Robert Haines [59]. For the same reason many UFOs observed by pilots do not cause a shock wave when bypassing despite their supersonic velocity.

(5) The sound of "blowing wind" is a characteristic signature of the SARG effect. It is clearly audible in Lab experiments. A similar sound is reported in a number of UFO cases when the observers are at some proper distance.

Chapter 5

(6) The SARG effect is more efficient if using a mixture of two gases: a working gas with a low atomic mass and a buffer gas with a larger breakdown voltage. Such a combination permits obtaining stronger LWs and probably better efficiency. Injection of preactivated plasma is expected to increase the power efficiency.

(7) The frequency of the AC field for plasma activation could be from a few kHz to a hundred kHz depending on the working gas and environment pressure. When it is in audible range it could be heard up to some distance. Some UFO observers report hearing a high pitch signal with changing frequency during the UFO take-off.

(8) The plasma activating electrodes must be designed with an appropriate curvature matching the LWs requirements and without sharp edges. A parasite corona in places away from the working electrodes may completely kill the SARG effect.

(9) The power requirements during landing and take-off of the spacecraft could be much larger than that required for maneuvering far from the Earth. This is so because a large amount of LWs is lost in the earth soil. This excess power may cause some imprinted marks on the ground. The landing – take-off area may contain unusual structural changes similar to the structural change of solid materials known as the Hutchison effect [37,38]. For a similar reason (dumping of emitted LWs), the spacecraft flying closer to the Earth's surface may unintentionally follow some power lines since they are elevated and may influence the plasma discharge.

(10) For a spacecraft at lower altitudes, the plasma envelope might be influenced by turbulence and especially the humidity that will cause an unstable motion. This type of motion known as hovering is a frequently observed feature of UFOs flying at low altitude.

(11) The technical realization of the propulsion system described in Chapter 4 is only one variant of a spacecraft with Field Propulsion. An alternative option is enclosing the plasma envelope in another external solid envelope of special material. In this case the efficiency of the SARG effect will be smaller but the spacecraft will have smoother motion. Such spacecraft could have more stable motion and will be more easily photographed and

detected by radar. It could not have a field shield protection and because of its vulnerability could be operated by a robot. The major advantage of such spacecraft is the possibility of using conventional EM sensors for monitoring the environment.

(12) The activated plasma provides a large reactive power that requires a specific approach for designing the HV AC source (more details about this are discussed in Chapter 4)

(13) Ordinary EM waves may not be suitable for communication between a spacecraft and the external world. Alternative communications with LWs could be used.

(14) Experiments with LWs must be done carefully since they obviously have a biological effect.

The author announced the discovery of the SARG effect at the 27 Annual conference of the Society for Scientific Exploration (SSE), 2008 [10]. The recorded talk at this conference was posted on the main website of SSE and on YouTube in 2009. In 2008 a patent application entitled "Method and Apparatus for Spacecraft Propulsion with a Field Shield Protection" was submitted to CIPO Canada (File No. 2,638,667 from 26 Aug 2008).

Appendix 1

APPENDIX 1 Ether drift experiments detecting the Earth motion in absolute space

Authors, Publ. date	Epoch	RA	δ	l	γ	V (km/s)	Type
David Miller, 1933	1925-26	4h54m	$-70^0 33'$	282^0	-35.2^0	208	Interferometer, continuous light
K. Illingworth - analyzed by Cahill & Kitto	1927					369+/-123	Interferometer, continuous light
S. Marinov, 1983	1973	Earth abs. velocity along instrument axis 130+/-100 km/s					Rotating mirrors, chopped light
S. Marinov, 1980	1975-76	13h23m +/- 20m	$-23+/-4^0$	313^0	38.9^0	303+/-20	Rotating mirrors, chopped light
R. Muller et al.	1976			Velocity towards Leo		~400	CMB
G. Smoot et al., 1977	1977	11h +/-0.6d	$6+/-10^0$	$245+/-15^0$	$54+/-10^0$	390+/-60	CMB
Wilkinson and Corey	1978?	12h+/-1h	$-21^0+/-21^0$	288^0	40^0	320+/-80	CMB
C. Monstein and J. Wesley, 1996	1978-96	$8.7^0 +/-3.5'$	$-1^0+/-10^0$	227.9^0	24.3^0	359+/-180	Muon flux anisotropy
S. Marinov, 1995	1984	12.5h +/-1h	$-24^0 +/-7^0$	397.5^0	38.4^0	362+/-40	Coupled shutters, chopped light
E. Silvertooth, 1986						378+/-?	Rotating mirrors, chopped light
M. Consoli et al., 2006		202^0	-44^0	309^0	18^0	276+/-71	Rotating optical resonators

Legend: Equatorial coordinates:
RA – Right Ascension
δ – Declination

Galactic Coordinates:
l – longitude
γ – latitude

V – velocity in Absolute space

Appendix 1

Ether drift experiments references:

1. D. C. Miller, The Ether-Drift Experiment and the Determination of the Absolute Motion of the Earth, Review of Modern Physics, **5**, 203-242, (1933)
2. G. F. Smoot et al., Detection of Anisotropy in the Cosmic Blackbody Radiation, Physical Review Letters, **39**, No 14, 898-901, (1977)
3. R. A. Muller, The Cosmic Background Radiation and the New Aether Drift, Scientific American, **238**, 64, (1978)
4. S. Marinov, Measurement of the Laboratory's Absolute velocity, General Relativity and Gravitation, **12**, 57-66, (1980)
5. S. Marinov, The interrupted "rotating disc" experiment, J. Phys A: Math. Gen, **16**, 1885-1888, (1983)
6. E. W. Silvertooth, Experimental detection of the ether, Speculations in Science and Technology, **10**, No 1, 3-7, (1986)
7. S. Marinov, New Measurement of the Earth's Absolute Velocity with the Help of the "Coupled Shutters" Experiment, (submitted by E. Schneeberger 10 years after Marinov's death), Progress in Physics, 1, 31-37, (2007).
8. C. Monstein and J. P. Wesley, Solar System Velocity from Muon Flux Anisotropy, Apeiron, **3**, No. 2, 33-37, (1996)
9. R. T. Chill and Kirsty Kitto, Michelson-Morley Experiments Revisited and the Cosmic Background Radiation Preferred Frame, Apeiron, **10**, No 2, 104-1017, (2003)
10. M. Consoli and E. Constanzo, Motion toward the Great Atractor from an ether-drift experiment, arXiv:astro-ph/0601420 v. 2, (2006) (analysis of 3 Modern laboratory experiments, two in Germany and one in Australia, based on active and cryogenic optical resonators).

APPENDIX 2. A Physical Model of the Electron According to the Basic Structures of Matter Hypothesis (article in Physics Essays*, v. 16, No 2, p. 180-195, 2003)

Stoyan Sarg

Abstract: A physical model of the electron is suggested according to the Basic Structures of Matter (BSM) hypothesis. BSM is based on an alternative concept about the physical vacuum assuming that the space contains underlying grid structure of nodes formed of super-dens sub-elementary particles, which are also involved in the structure of the elementary particles. The proposed grid structure is formed of vibrating nodes possessing quantum features and energy well. It is admitted that this hypothetical structure could be accounted for the missing "dark matter" in the Universe. The signature of such "dark matter" is apparent in the galactic rotational curves and in the relation between masses of the supermassive black whole in the galactic centre and the host galaxy. The suggested model of the electron possesses oscillation features with anomalous magnetic moment and embedded signatures of the Compton wavelength and the fine structure constant. The analysis of the interactions between the oscillating electron and the nodes of the vacuum grid structure allows obtaining physical meaning for some fundamental constants.

Keywords: physical vacuum, structure of the electron, fine structure constant, Compton wavelength, anomalous magnetic moment, dark matter, Planck frequency, unified theories

1. Introduction

The "dark matter" is a hot topic in the cosmology today. Currently it is accepted that the "dark matter" predominates the visible matter in the Universe. In recent years it has been found that most of galaxies contain in their centre a supermassive black hole in order of billion solar masses. A surprising strong relation has been found between the mass of the supermassive black hole and the total mass of the whole galaxy, so they are in kind of balance[1] (L. Ferrarese, D. Merrit, 2000). Another peculiar fact for existence of hidden matter comes from the rotational curves of the galaxies. One of the largest

* International Journal Dedicated to Fundamental Questions in Physics

APPENDIX 2

rotation curve database of spiral galaxies clearly shows that the "dark matter" is rather a rule, than exception (see the article "An analysis of 900 optical rotation curves: Dark matter in a corner?", by D. F. Roscoe, (1999)[2]). It stands to reason raising a question: Isn't the "dark" matter a hidden type of matter around us and even "within us"? Such idea further leads to the conclusion that the currently adopted concept about the physical vacuum may not be correct. This required an extensive study of some features of the physical vacuum such as the Zero Point Energy, the quantum fluctuations, the vacuum polarization, the Plank's length and frequency and so on. In such aspect, the theoretical articles provided by T. H. Boyer[3], H. E. Puthoff[4,5,6], H. E. Puthoff et al[7], B. Haisch et al.[8] appeared quite useful. F. M. Meno[9] envisions hypothetical three-dimensional non-spherical particles called gyrons possessing a gyroscopic effect. He associate the Planck's length and mass to some of the gyron's parameters, although he does not suggest a detailed physical model of this gyron and does not envision a possible organization of the gyrons into stable structures. The articles "Experimental evidence that the gravitational constant varies with the orientation" by M. M. Gershteyn et al[10] and the "Speed of gravity revisited" by M. Ibison et al[11] lead to the idea that the Newton's law of gravitation might be derivable instead of postulated. This idea obtained some theoretical treatment by H. E. Puthoff[4] (1989) who derived the Newton's law of gravitation starting from the Planck's frequency, ω_{PL}, and using one hypothesis of Sakharov.

$$\omega_{PL} = [2\pi c^5 / hG]^{1/2} \qquad (1)$$

where: c – is the light velocity, h – is the Planck constant, G – is the gravitational constant

Here we may express an idea about existence of some hypothetical structure in the microscale range, related in some way to ω_{PL}. This could be regarded as a further development of the concept of the string theories, which assume an existence of some hypothetical string-like objects (open or closed loops) in a microscale range possessing a finite length but zero thickness. What could be the results if these hypothetical extended objects possessed a finite width, while their dimensions are far beyond the observational limit. In such case these strings should be regarded as material objects in a three-dimensional space and they might be organized in structures. It stands to reason that we are able to

APPENDIX 2

observe enormously large structures in the macroscale range of the Universe, but structures may exist also in the microscale range[12].

One additional consideration that the Newton's law of gravitation might be derivable from a more fundamental one comes from its comparison with the law of optical radiation. In its simplest form, when the surface of two areas A_1 and A_2 are parallel each other, the irradiation flux, Φ, is given by: $\Phi = LA_1A_2/r^2$ where L – is the emitted radiance and r – is the distance between the two surfaces (visible in the subtended angle). If the two bodies are parallel disks, the radiation law depends only on the visible surfaces but not on the disk thickness. In the same time the Newton's gravitational law depends on the thickness or the bulk matter of the bodies. But why they both have one and a same dependence on the distance? It seams that the Newton's gravitational mass could have some dependence on the area of the closed surface of some unknown real structure on which some hypothetical substance may exercise pressure.

The above-mentioned citations and logical considerations were helpful in the search for appropriate model of the alternative vacuum concept. An idea was born that the Planck's frequency could be a parameter of some intrinsic type of matter involved in some unknown sub-elementary particles from which both - the vacuum structure and the elementary particles are built. These hypothetical sub-elementary particles may possess enormous mass density and may interact between themselves in a classical void space. Their gravitational interactions, however, may distinguish from the Newton's gravitation by the degree of proportionality to the distance. In such aspect, we me refer such type of gravitational interactions as Intrinsic Gravitation (IG). The hypothetical sub-elementary particles, for instance, may form stable structures if their IG in a classical void space is inverse proportional to the cube of the distance. In such way they may form a stable spatial grid. **In the same time the Newton's gravitation acting between the elementary particles and their formations (atomic nuclei, atoms, molecules) could be a result of the IG field propagation trough the interconnected elements of the spatial grid.** The IG gravitation, however, may leak at some close distance between atoms and molecules (some types of Wan der Waals forces) or well polished solid objects (Casimir forces).

APPENDIX 2

2. Brief introduction into the concept of the BSM hypothesis

The presented above considerations served as a starting point for development of a hypothesis called Basic Structures of Matter (BSM)[13], associated to the class of the unified theories. According to the BSM concept[14] the intrinsic gravitation (IG) force, F_{IG}, between two objects comprised of same type of intrinsic matter put in a classical void space is proportional to the product of their intrinsic masses and the intrinsic gravitational constant and inverse proportional to the cube of the distance.

$$F_{IG} = G_0 \frac{m_{01} m_{02}}{r^3} \qquad (2)$$

where: G_0 – is the IG constant, m_{01} and m_{02} – are intrinsic masses, r – is the distance.

It is assumed that the space known as a physical vacuum possesses a underlying grid structure formed of two types of sub-elementary particles arranged in nodes. These two sub-elementary particles are built respectively by two types of intrinsic matter with different density. They both have a shape of hexagonal prisms with length to diameter ration > 1, while the dimensional ratio between both prisms is 2:3. They possess also a similar internal structure with twisted component, but left and right handed respectively. Prisms of the same type (intrinsic matter and handedness) are attracted in a pure void space by forces according to defined above IG law. The attraction forces between the different types of prisms, however, are smaller and dependable on the node distance and they may convert to repulsion at some critical value of this distance. Additionally the prisms of both types possess IG anisotropy along their axis with a left and right twisting component respectively, defined by their lower level structure. For this reason they are called twisted prisms, although, they are not externally twisted. The formation of such sub-elementary particles is possible from much simpler spherical particles, following pure geometrical principles and preservation of the integrity of the lower level structures in the upper level structures. A hypothetical scenario for this is provided in Chapter 12 of BSM. According to the BSM concept, the two types of prisms build the underlying structure of the physical vacuum and the elementary particles as well. The structural integrity in both cases is assured by the IG law, defined by Eq. (2). The elementary component of the vacuum structure is a node called a Cosmic Lattice (CL) node. The CL node is formed of four prisms of a same type held by IG forces in positions like the four axes in a

APPENDIX 2

tetrahedron, but the connected prisms have some limited freedom of angular deviation. The vacuum structure is formed by alternatively arranged nodes of both types with some gaps between the prisms of the neighboring nodes. The spatial CL structure is similar to the atomic lattice in a diamond. It is assumed that such structure fills the volume of the visible Universe, so the space in BSM is referenced as a CL space. The elementary particles are built by the same prisms, but arranged in configuration of helical structures inside of which a different type of spatial structure (internal lattice) from the same prisms exists. The internal lattice, however, is denser than the CL structure, so the latter could not penetrate inside the internal lattice. Therefore, the CL space should exercise a pressure on the internal lattice of the particle. The pressure parameter of the CL space leads to derivation of a mass equation in BSM (Chapter 3). It is estimated that the node distance is in order of 10^{-20} (m), while the overall size of any elementary particle is larger by few orders. In the same time the density of the intrinsic matter from which the prisms are built is many orders larger that the average density of any elementary particles. In such conditions the CL space is able to carry the elementary particles, while an accumulation of these particles in a closed volume may influence but very weakly the node distance of the CL space in a close proximity (a large mass accumulation may distort slightly the node distance in the surrounding space leading to a space curvature according to the General Relativity).

One specific feature of the CL space is the ability of the CL nodes to be displaced by the denser internal lattice of the moving elementary particles (every particle is in motion due to the galactic rotation). Such effect involves a disconnection, a displacement with simultaneously folding of the CL node and returning, unfolding and reconnection to the previous position of CL structure. The connection energy during the displacement is transferred to a kinetic energy. Such unique feature does not have counterpart in any concept of aether or ideal fluid. The folding properties of the CL nodes are also closely related to the inertial properties of the atomic matter in CL space and play a role in the equivalence between the gravitational and inertial mass.

Analysing the dynamics and mutual interactions of the CL nodes, it is possible to associate some of their features with known physical parameters and constants such as, the Zero Point Energy of the vacuum, the light velocity, the Compton frequency (or wavelength), the permeability and permitivity of the free space.

APPENDIX 2

Figure 1 illustrates the geometry of a single CL node in a position of geometrical equilibrium. The four prisms are held by IG forces defined by Eq. (2).

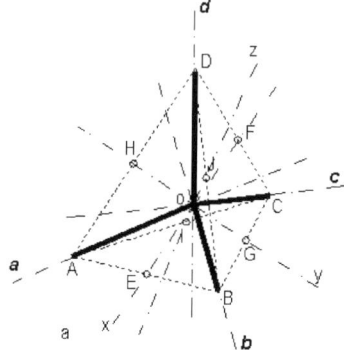

Fig. 1. CL node formed of four prisms shown by thick black lines

The CL node is characterized by two sets of axes: one set of 4 axes along anyone of the prisms called *abcd* axes, and another set of 3 orthogonal axes called *xyz* axes. In a geometrical equilibrium the angle between anyone of *abcd* axes is 109.5°. The external tips of the prisms in a geometrical equilibrium define the apex points of a tetrahedron. The *xyz* axes pass through the middle of every two opposite edges of the tetrahedron. In the same time, the orthogonal *xyz* axes of the neighbouring CL nodes are commonly aligned. Such arrangement assures complex individual oscillations of the CL node, from one side and strong interactions between the neighbouring nodes, from the other. The dynamics and interactions are both governed by the IG law acting between the two intrinsic matter substances of the prisms and the time constant for this matter.

The dynamical behaviour of the CL node is studied by estimating the shape of the return forces (under condition of IG law) acting on the deviated from the central position CL node and keeping in mind that the node geometry is flexible. For simplification of the analysis, the neighbouring four CL nodes are considered stationary, while their interconnecting prisms are always aligned to the CL node under consideration (for details see BSM monograph, BSM_appendix2-1.pdf). This also means, that the IG interactions propagate faster than the oscillation period of the CL node. The shapes of the return forces along anyone of *xyz* and *abcd* axes are shown respectively in Fig 2. (a) and (b). Two symmetrical minimums appear along anyone of *xyz* axes and one minimum along the positive direction of anyone of *abcd* axes. From a point of view

APPENDIX 2

of CL node dynamics, they could be associated with energy wells, responsible for the ZPE (zero point energy) of the vacuum.

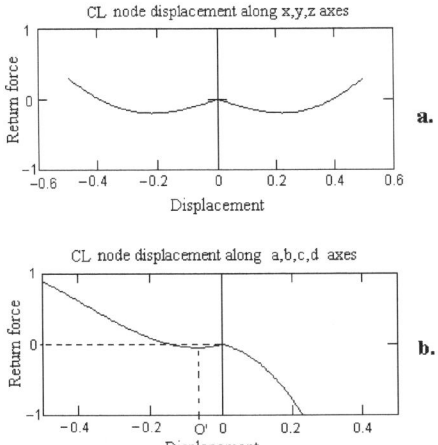

Fig. 2. Return forces versus displacement of the CL node along one of *xyz* axes (a) and *abcd* axes (b). Both scales are in relative units.

The shape and the different stiffness of the return forces along *xyz* and *abcd* axes indicates that the CL node will possess a complex type of oscillations in which two types of cycles are identifiable: a proper resonance cycle and a SPM cycle (the latter is described by a Spatial Precession Momentum vector). The trace of the proper resonance cycle is approximately flat but open curve with four bumps, as shown in Fig. 3.

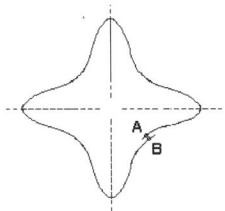

Fig. 3. Trace of single proper resonance cycle of the CL node

The bumps of the trace curve centred on the two orthogonal axes are caused by the different stiffness for node deviations along *abcd* and *xyz* axes. The points A and B from the resonance cycle are pretty close but not coinciding, so the segment AB points almost at 90 deg in respect to the drawing plane. The lack of coincidence between any initial (A) and final (B) point for one proper resonance cycle is a result of the spatial positions of the return forces minimums along the two set of axes. The CL node dynamics for the proper resonance cycle could be described by a vector called a Node Resonance Momentum (NRM).

The average plane of the trace is slightly rotating with every NRM cycle, so after a large number of such cycles the node trace will passes through the same (arbitrary selected) initial point A. This second type of cycle is called a SPM cycle. The vector describing this cycle is called a Spatial Precession Momentum vector (SPM). The number of the resonance cycles in one SPM cycle, estimated in BSM, is quite large but constant (due to the mutual interactions of the oscillating CL nodes). The analysis in BSM indicates that this number is related to the magnetic permeability of free space (section 2.11.3 in chapter 2 of BSM).

The tip of SPM vector for one full cycle circumscribes a closed surface with a central point of symmetry and six bumps along the axes *xyz*. Such type of surface is referenced in BSM as a SPM quasisphere. It is found that if the resonance cycle of the CL node is related to the energy wave propagation with a light velocity, the SPM cycle should be related to a particular quantum feature of the CL space that assures the constant value of the light velocity. This is explainable by the quantum properties of the SPM quasispheres and their mutual interactions. The light velocity is considered as energy momentum propagation between two neighbouring nodes (considering *xyz* interconnection coordinates) for one resonance cycle of the CL node (section 2.11 in Chapter 2 of BSM). The frequency of SPM cycle is associated to the well-known Compton frequency. In absence of any electrical charge, the SPM quasisphere possesses a central point of symmetry. It is called a Magnetic Quasisphere (MQ), because it could provide a physical meaning of the magnetic line. The magnetic line could be formulated as **a closed loop in CL space involving only MQ type of nodes whose SPM frequencies are synchronized by a running phase propagating with a light velocity.** Such spatial configuration may exhibit features allowing explanation of the stability and direction of the magnetic line, for example:

- The CL nodes of right-handed prisms are commonly synchronized
- The CL nodes of the left-handed prisms are commonly synchronized
- The phase difference between the involved left and right handed nodes determines the direction of the magnetic line,

APPENDIX 2

referenced to the laboratory frame, for example, +90 deg phase difference for N-S direction and -90 deg phase difference for S-N direction.

- The involved MQ nodes may additionally have a helical arrangement along the closed loop.

The above considerations are for permanent magnetic field. In case of alternative magnetic field, the commonly spatially dependable synchronizations of the left and right-handed nodes vary with the time.

Aligned MQs with a spontaneous phase synchronization (with light velocity) may also exist in an open loop, but temporally. This is a normal state of the oscillating CL node when considering the mutual interactions of the neighbouring CL nodes and this effect appears to be related to the magnetic permeability of the free space.

In a presence of charge particle, the SPM quasisphere obtains a deformation as an elongation along its diameter connecting two opposite bumps, so it is called an Electrical Quasisphere (EQ). The shapes of MQ and EQ are shown in Fig. 4.

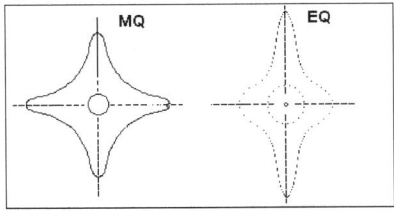

Fig. 4 Shape of MQ (left) and EQ (right)

For analysis simplification when studying the dynamics, the positions of the CL nodes could be considered as stationary in a laboratory frame. The electrical field could be presented as spatially oriented and synchronized EQ CL nodes. When studying the conditions of energy propagation as a wave, it is convenient to use imaginary *running CL nodes*. Then the phase propagation of the SPM vector with a speed of light through stationary positioned CL nodes can be regarded as a running SPM vector. In this manner, the temporal variation of the common synchronization of the CL nodes is easily studied. The analysis in such approach leads to unveiling the structure of the photon. It is found that the EQ type node possesses a larger energy than MQ type (see section 2.10.4.3 Chapter 2 of BSM). The photon wavetrain can be presented as a complex arrangement of *running EQs* with a decreasing elongation from the central axis of the wavetrain to its boundary radius, where they are converted to *running MQs*. Thus, it appears that the photon wavetrain possesses boundary conditions (a long standing problem). In the same time the running EQs are align in a helix with a step equal to the photon wavelength.

The analysis of the CL node dynamics as EQ and MQ type and the suggested photon wavetrain structure in a normal CL space (possessing a normal Zero Point Energy) are presented in Chapter 2 of BSM. The CL space with a subnormal Zero Point Energy and the behaviour of the charge particles in such case are analysed in Chapter 4 of BSM.

The applied new approach allows admitting that the elementary particles also possess underlying structure built by the same sub-elementary particles – the two types of prisms. BSM analysis leads to a conclusion that the stable particles, such as proton, neutron and electron (and positron) possess stable structures with well-defined spatial geometry and denser internal lattices. They are comprised of complex but understandable three-dimensional helical structures whose elementary building blocks are the mentioned above prisms, arranged in a strong particular order. The analysis provided in chapter 8 of BSM leads to a conclusion that the protons and neutrons are spatially arranged in the atomic nuclei[15]. If the suggested vacuum structure is real, the interpretation of the scattering experiments should be reconsidered, because both, the vacuum structure and the structure of the elementary particles have not been taken into account, so far.

3. A physical model of the electron built by the suggested sub-elementary particles

According to the BSM concept, the electron possesses the simplest structure among the stable elementary particles. The suggested physical model of the electron is comprised of three helical structures, one inside another, as illustrated in Fig. 5. The helical structure is comprised of a helical envelope and internal lattice inside this envelope. All of them are built by the suggested sub-elementary particles (prisms). The axial section of an elementary core from any helix envelope contains 7 prisms of the same type, one in the centre and 6 in the periphery. In the same time they are axially displaced (as shown in Fig. 5), so the helix could be considered as formed of stacked elementary cores. The two helical structures of the electron possess denser internal lattices located in the internal spaces of the helix envelopes (not shown in this figure).

APPENDIX 2

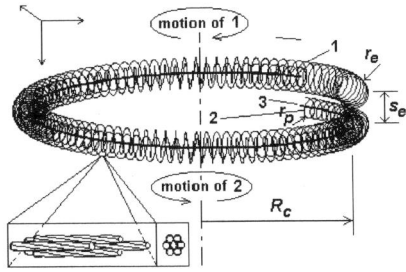

Fig. 5. Oscillating electron is comprised of three helical structures: 1 – external negative, 2 - internal positive, 3 – internal negative core. The internal lattices are not shown. The expanding box in the lower left side shows an elementary node of the helical structure 1, formed of 7 right-handed prisms (they are not externally twisted, the twisting is for a concept visualization only)

The dimensions of the physical components of the electron structure are denoted as : R_c – the Compton radius of electron (known), r_e - a small electron radius, r_p - a small positron radius, s_e – a helical step. The derivation of these dimensions is discussed later.

The electron structure, shown in Fig. 5 has two internal lattices spaced inside the volume enclosed by the two helical structures. Each one is built of same type of prisms like their envelopes. The outer lattice has a larger whole it its radial section where the internal first order helical structure oscillates. The other internal lattice has a smaller whole where the internal core oscillates.

The geometrical considerations allowing building of internal lattice are illustrated by Fig. 6.

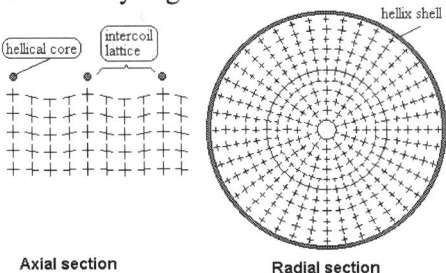

Fig. 6 Configuration of the internal lattice of type RL (Rectangular Lattice) inside the cylindrical space enveloped by the helical core, which forms the helical structure. The actual number of layers in the radial section is much larger than this shown in the figure, because the prism size is a few orders smaller than the radial section diameter

APPENDIX 2

Every RL node is comprised of six prisms of the same type. The axial section contains number of concentric layers. Starting from the cylindrical boundary defined by the helix envelope, the most external layer is connected to the helix by IG forces, while every internal layer is connected to the neighbouring external one. The thickness of every internal layer is half of the thickness of the neighbouring external layer. The radially aligned prisms of the neighbouring nodes are without gaps, while the gap length between the tangentially aligned prisms in the radial section varies when moving from external to the internal radius of the layer.

When considering an open formation of helical structures, as for the electron (the both ends are not connected as in a torus) the overall configuration could not be stable if the internal lattices are of rectangular type. Such formation, however, can be stabilized if the internal RL structures get some twisting.

Figure 7 illustrates the radial section of untwisted (a) and twisted (b) RL structures, referenced respectively as RL and RL(T).

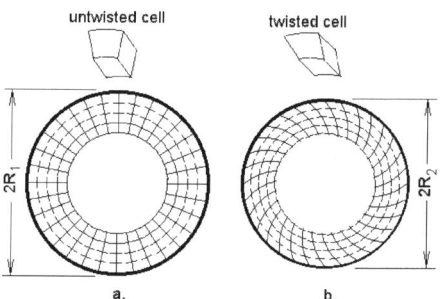

Fig. 7. Radial section of untwisted (a) and twisted (b) RL structures, referenced respectively as RL and RL(T). $R_2 < R_1$

The stiffness of the RL structure defined by the prism density is about 1000 times larger than the stiffness of the CL structure of the vacuum, so the volume of the RL structure is not penetrative even for folded CL nodes. Consequently it displaces the CL structure, or in other words, it feels a CL pressure. This is a Static CL pressure.

The twisted radial stripes of RL(T) modulate the dynamical properties of the CL nodes in the surrounding space, more accurately their SPM quasispheres. In such way, they become EQ type nodes arranged in line extensions from the twisted radial stripes of RL(T). These spatially arranged EQ nodes form the electrical field of the charge particle, in our case – the electron. This is illustrated in Fig. 8. It is evident that in a proximity range the electrical lines might be slightly curved but in a far range they appear as emerging from a point.

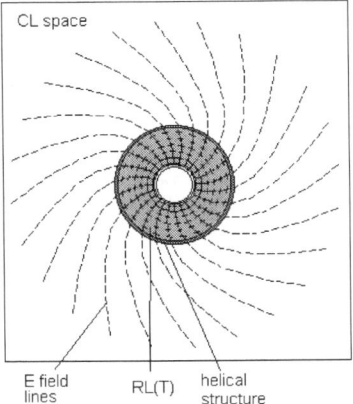

Fig. 8 Proximity E-field lines (in CL space) emerging from the RL(T) structure

One from both types of the prisms (for instance the right handed) could be associated with the negative electricity while the other with the positive one, but keeping in mind that the electrical charge is a property of the CL space, related to the presence of EQs and not a property of the prism itself.

The external helical structure of the electron, referenced in BSM as an external shell, possesses an internal denser lattice (from right handed prisms, for example). It is responsible for creation of EQ type CL node as radial extensions from the RL(T). The curved line extensions in the proximity to the external electron shell are of essential importance for the confined motion that the electron exhibits in CL space.

The internal helical structure with an internal RL(T) (from a left-handed prisms, respectively) with a central core (from right-handed prisms) is an **internal positron**. When completely inside in the external electron shell, it is not able to modulate the external CL space, but when it is outside it appears as a positive charge. When the internal positron oscillates inside the electron external shell, its charge only partially appears in CL space with a rate of the oscillation cycle. Due to the high oscillation frequency (discussed below) only its magnetic signature may interact with the external CL space.

APPENDIX 2

Considering the oscillation properties of the suggested model of the electron it could be regarded as a three-body system: an external helical structure with its internal lattice (external shell built of negative prisms), an internal helical structure with its internal lattice (internal shell built of positive prisms) and the central core (built of negative prisms). Both, the internal helical structure and the central core oscillate in conditions of ideal bearing because their central positions are kept by IG field and the whole structure has a complete helical symmetry in respect to the central core. In such conditions the electron structure will have two proper frequencies

- a first proper frequency: for the oscillations between the external electron shell and the internal positron
- a second proper frequency: for the oscillations between the internal positive shell and the central negative core (a proper frequency for the internal positron).

From the analysis of the dynamical properties of the suggested structure it appears that the first proper frequency of the electron is equal to the SPM frequency of the CL node. This is the well-known Compton frequency.

4. Quantum motion of the electron and derivation of its structural parameters

The value of the physical constants and parameters used in the presented analysis are given in Table I.

Table I. Used fundamental constants[16] according to CODATA 98

Constants Name	Value		Unit
α	$7.297352533(27) \times 10^{-3}$		fine structure constant
c	2.99792458×10^{8}	m/s	light velocity
λ_c	$2.426310215 \times 10^{-12}$	m	Compton wavelength
h	$6.62606876(52) \times 10^{-34}$	Js	Planck constant
ε_0	$8.854187817 \times 10^{-12}$	F/m	permitivity of free space
m_e	$9.10938188(72) \times 10^{-31}$	kg	electron mass
a_0	$0.5291772083(19) \times 10^{-10}$	m	Bohr radius
K_J	$483597.898(19) \times 10^{9}$	Hz/V	Josephson constant
R_∞	$1.973731568549(83) \times 10^{7}$	1/m	Rydberg constant

APPENDIX 2

It is assumed and extensively discussed in BSM hypothesis that the prisms, formed of superdense intrinsic matter possess quite different inertial properties in a pure void space. (The very high interaction frequency of this matter may be closer to the Planck frequency and consequently it may have a very small inertial property). It is apparent that the CL structure from its side possesses a time constant which is obviously defined by the proper resonance frequency of the CL node.

The analysis of the motion behaviour of the electron structure in CL space leads to a conclusion that it will possess a preferable type of a screw-like motion. Such motion in the CL space environments is possible if some CL nodes are temporally disconnected, displaced and then returned and reconnected to the CL space.

Such type of motion is referenced as a confine one. Two types of confined motion are identified: (1) a confined motion with optimal and sub-optimal velocities; (2) a confined motion with super-optimal velocities

4.1 Confined motion with optimal and sub-optimal velocities.

Both, the CL node and the rotating electron oscillate with a Compton frequency. It is found that when the tangential velocity of the rotating and oscillating electron is equal to the light velocity, the phase of its first proper frequency matches the phase of the SPM vector, propagating with a light velocity. In the same time, the internal core oscillation (with a proper frequency of three times the Compton frequency) provides a third harmonic feature for this motion. As a result the rotating and oscillating electron exhibits a maximum interaction with the CL space - a kind of quantum interaction. The electron axial velocity for this case is $V_{ax} = \alpha c$ (corresponding to a kinetic energy of 13.6 eV). It is referenced in BSM as an **optimal confined velocity** and the motion respectively as an **optimal confined motion**. We may consider that any point of the electron structure corresponding to a radius R (measured from the central point of the whole structure) moves with a tangential velocity equal to the speed of light (because it appears from the analysis that $r_e \ll R$). For such point of the structure the following relations are valid:

Peripheral velocity: c - path: $(4\pi^2 R^2 + s_e^2)^{1/2}$

Axial velocity: V_{ax} - path: s_e

APPENDIX 2

Then the axial velocity is:

$$V_{ax} = cs_e / (4\pi^2 R^2 + s_e^2)^{1/2} \tag{3}$$

From the Bohr model of hydrogen we know that the kinetic energy of 13.6 eV corresponds to an electron motion in orbit of radius a_0, with a velocity given by Eq. (4).

$$V_{ax} = (q_0^2 / 2h\varepsilon_0) = \alpha c = 2.187691 \times 10^6 \quad \text{m/s} \tag{4}$$

where: q_0 – is the elementary charge, h – is the Planck constant, ε_0 - is the permitivity of the vacuum, α - is the fine structure constant and c – is the light velocity.

Therefore, we arrive to the following two conclusions:

(1) The screw-like motion of the suggested electron model with a tangential velocity equal to the speed of light is energetically equivalent to an electron motion in a circular orbit of radius, a_0, according to the Bohr model of the hydrogen.

(2) The fine structure constant appears to be a ratio between the axial and tangential velocity of the electron, when it performs an optimal confined motion.

Combining Eq. (3) and (4) we obtain a step to radius ratio of the electron

$$R/s_e = (1-\alpha^2)^{1/2} / 2\pi\alpha = 21.809 \tag{5}$$

Now let assuming that the first proper frequency of the electron is equal to the Compton frequency (a parameter of the CL node) and the electron structure makes one full rotation for time duration equal to the Compton time, t_c that is a reciprocal to the Compton frequency.

$$path = 2\pi R = ct_c = c(1/v_c) \tag{6}$$

Solving the system of Eq. (5) and (6) we get the value of R and s_e.

$R = 3.86159 \times 10^{-13}$ (m) – the large radius of electron

$s_e = 1.77061 \times 10^{-14}$ (m) – the helical step

It is not a surprise that the obtained value for R is exactly the Compton radius R_c, which is experimentally determined by Arthur Compton. Substituting R with R_c in Eq. (5) and having in mind that $2\pi R_c = \lambda_c$, we obtain an expression for the helical step, s_e.

$$s_e = \alpha \lambda_c / (1-\alpha^2)^{1/2} \tag{7}$$

APPENDIX 2

The Compton wavelength, λ_c is related to the Compton frequency, v_c, by the simple expression $\lambda_c = c/v_c$. The light velocity is related to the resonance frequency of the CL node, while the Compton frequency is the SPM frequency. Then from Eq. (7) follows a conclusion that:

The suggested model of the electron is characterised by two embedded fundamental constants: the fine structure constant and the Compton wavelength.

From number of considerations given in section 3.6 and 3.11.2 of Chapter 3 in BSM it appears that $s_e \approx 2r_e$, and it is assumed that this relation is more accurately expressed by the gyromagnetic factor, g_e, that is experimentally determined with high accuracy.

$$s_e = g_e r_e = 2.002319 r_e \qquad (8)$$

From the analysis of the Fractional Quantum Hall experiments in Chapter 4 of BSM, it is found that: $r_p/r_e = 2/3$. Then all geometrical parameters of the electron are determined.

At the optimal confined motion with a velocity $V_{ax} = \alpha c$ (13.6 eV) the quantum interaction of the oscillating electron with the oscillating CL nodes are strongest. It is apparent that the electron may perform a screw-like motion also with smaller velocities. Let considering these velocities for which the electron makes a complete rotation for a whole number of first proper frequency cycles. Such set of axial velocities could be expressed by Eq. (9), where n is an integer.

$$V_{ax} = \alpha c/n \qquad (9)$$

If using the kinetic energy of the electron instead of its axial velocity we have

$$E = 0.5 h v_c \alpha^2 / n^2 \quad \text{(J)} \qquad (10)$$

$$E_{ev} = 0.5 h v_c \alpha^2 / (n^2 q_0) \quad \text{eV} \qquad (11)$$

where: h – is the Planck constant, v_c - is the Compton frequency, q_0 - is the electron charge

The integer n is called in BSM a **subharmonic number**, in order to notify the quantum motion conditions of the electron. A quantum motion with a first harmonic velocity corresponds to 13.6 eV, with a second subharmonic - 3.4 eV, with a third subharmonic - 1.51 eV and so on. It is evident that the introduced subharmonic

APPENDIX 2

number, n, matches the quantum number of the electron orbit in the Bohr atomic model. In the same time it is informative about the rotational spin motion of the oscillating electron if referencing its rotation cycle to the SPM cycle of the CL space:

    13. 6 eV - 1 rotation cycle per SPM cycle (an optimal confined motion)
    3.4 eV - 1/2 rotation cycle per SPM cycle
    1.51 eV - 1/3 rotation cycle per SPM cycle
    0.85 eV = 1/4 rotation cycle per SPM cycle
    SPM cycle period = Compton time

    Analysing the confined motion of the electron, it is possible to get some insight about its influence on the SPM quasispheres surrounding its trace of motion. It is found that the surrounding EQs of the moving electron will cause a formation of spatially ordered synchronization of the surrounding MQs in closed loops, i. e. creation of magnetic lines. In such aspect it is useful to analyse the magnetic radius of the electron at different subharmonic numbers. We may consider that the rotating IG field of the internal lattice of the electron helical structure (that modulates the CL space) could not exceed the light velocity. Then the magnetic influence could be extended up to some limited range and we may regard it as a magnetic radius.

    The magnetic radius of electron with a kinetic energy of 13.6 eV is obtained from the analysis of the quantum magnetic field Φ_0 (see section 3.11 in Chapter 3 of BSM): $\Phi_0 = h/q_0$, where h – is the Planck constant, q_0 – is the electrical charge. The obtained value of r_{mb} for 13.6 eV is almost equal to R_c, but slightly larger due to the finite thickness of the electron helical structure. The electron model gives also some insight about its magnetic moment. The magnetic moment of the electron is considered anomalous because it is distinguished from the Bohr definition of magnetic moment by the term $\alpha/(2\pi)$.

$$\mu_e = \frac{q_0 h}{4\pi m_e}(1+\frac{\alpha}{2\pi}) \qquad (12)$$

where: m_e – is the mass of the electron

    The anomalous term in Eq. (12) appears because the overall shape of the electron is not a torus but a single coil, possessing a helical step. Having such shape the electron is able to advance by a size of a full step, s_e, for one revolution, so these feature contributes

APPENDIX 2

to the "anomalous" term $\alpha/(2\pi)$. This feature is not taken into account when the magnetic moment is defined from the considerations of the Bohr atom. The magnetic moment is discussed in details in section 3.11 Chapter 3 of BSM.

4.2 An electron motion with super-optimal velocities

The optimal confined motion of the electron (axial velocity of $V_{ax} = \alpha c = 2.18769 \times 10^6$ (m/s)) could be regarded as an ideal case of the screw-like motion. In such motion the rim of the electron structure slides like in a thread and the oscillation of the central core (with a proper frequency three times higher than the first proper frequency) provides a hummer-drill effect enhancing the interaction with the stationary CL nodes. Keeping in mind that the phase of the SPM frequency propagates with the speed of light it is evident that the screwing electron is moving as in a lubricated thread. At this quantum velocity the electron exhibits a maximum quantum interaction with the CL space. For larger velocities (or energies larger than 13. 6 eV), the motion is still confined, but the screwing is not like in a thread (because no point of the electron structure could exceed the light velocity, which is restricted by the proper resonance frequency of the CL node). Therefore, we may expect a decrease in the quantum efficiency for such velocities. This is discussed in section 3.11.A.1, Chapter 3 of BSM, where an expression for the quantum efficiency is derived. The obtained expression appears to be a reciprocal function of the relativistic gamma factor. This is in agreement with the mass increase of the electron at relativistic velocities.

5. Rydberg constant as a signature of the optimal confined motion of the electron

Let considering a quantum motion of the electron (13.6 eV) with an optimal confined velocity ($n = 1$, n – is a subharmonic number). The electron energy for $n = 1$ according to Eq. (10) is

$$E = 0.5 h v_c \alpha^2 \quad \text{(J)} \qquad (13)$$

The energy of 13.6 eV photon is expressed by

$$E = h v_c = hc\sigma \quad \text{(J)} \qquad (14)$$

where: $\sigma = 1/\lambda_c$ - is the wavenumber, v_c - is the Compton frequency, c – is the light velocity

APPENDIX 2

Equations (13) and (14) provide one and a same energy (13.6 eV). Solving this system for σ, we get the Rydberg constant in wavenumbers

$$\sigma = R_\infty = \frac{v_c \alpha^2}{2c} = 1.097373156 \times 10^7 \quad [1/m] \quad (15)$$

The suggested model of the electron contains an embedded fine structure constant as seen from Eq. (7). Additional analysis in BSM (section 2.9.6.B of Chapter 2 and section 9.7.5 of Chapter 9, from the first edition of BSM) indicates that the fine structure constant is in fact an intrinsic parameter of the CL space. The Compton frequency is also a CL space parameter characterizing the CL node dynamics. Then from Eq. (15) it follows that the Rydberg constant is also a CL space parameter. The way it was derived indicates that **the Rydberg constant appears as a characteristic feature of the quantum motion of the electron with an optimal confined velocity.**

6. Quantum motion of the electron in a closed loop trajectory.

The orbital motion of the electron in atoms could be regarded as a motion in a closed loop, whose trajectory follows the equipotential surface of an electrical field defined by one or more positive charges.

Let considering a repetitive motion in a closed loop. The modulation properties of the internal RL(T) lattice in a repetitive motion may cause distortion of the MQs (that is a normal state of the SPM vector) converting them into EQs. This will affect the orbital conditions defined by the proximity field of the proton. Let assuming that the orbital motion of the oscillating electron tends to adjust itself to this change by exchanging some reactive energy with the CL space, that is hidden for the external observer. Then we may analyse the phase repetitions of the two proper frequencies of the electron and the conditions of their match to the phase of the SPM frequency of the CL nodes. In such way we may assume that the stability of a repetitive motion in such loop will depend on the phase repetition for both, the first and the second proper frequencies of the electron.

We will try to find the smallest path length at which the quantum loop conditions for an electron moving with a velocity corresponding to $n = 1$ (13.6 eV) is fulfilled. **Initially we will ignore the relativistic effect for simplicity.** It is reasonable to look

APPENDIX 2

for a path length defined by some CL space parameter. One such parameter is the Compton wavelength λ_c, related to the Compton frequency v_c by the simple expression $\lambda_c = c/v_c$. For one orbital cycle in a closed loop with length λ_c, the number of turns (electron structure rotations), N_T, is:

$$N_T = \lambda_c / s_e = 137.03235 \qquad (16)$$

The value of N_T could be regarded as a condition for a phase repetition for two consecutive passages through a chosen point in the loop, keeping in mind a confined (screw-like motion) of the electron. The trace length of $\lambda_c = 2.4263 \times 10^{-12}$ (m), however, is quite small, when comparing to the Bohr orbit length of $2\pi a_0 = 3.325 \times 10^{-10}$ (m). Therefore, we may look for a phase repetition conditions at larger loop length. From Eq. (16) we see that N_T is close to $1/\alpha = 137.036$ and this seams not occasional. Then, we may substitute N_T in Eq. (16) by $1/\alpha$ and multiply the result by λ_c. The latter is a CL space parameter from one side (a length of SPM phase propagation for one SPM cycle) and from the other - the circumference length of the electron structure. In such case we obtain:

$$N_T \lambda_c = \frac{1}{\alpha} \lambda_c = 3.24918460 \times 10^{-10} \quad (m) \qquad (17)$$

We see that the obtained value of Eq. (17) having a dimension of length is equal to the Bohr orbit length given by CODATA 98 (see Table 2) up to the 9th significant digit.

$$2\pi a_0 = 3.24918460 \times 10^{-10} \quad (m) \quad \text{CODATA 98} \quad (18)$$

where: $a_0 = 0.5291772083 \times 10^{-10}$ (m) - is the radius of the Bohr atomic model of hydrogen.

The expression (17) is not something new. The important, fact, however, is the way of its derivation related with the suggested physical model of the electron. The obtained loop length appears equal to the orbit length of the Bohr atom, defined by Bohr atomic radius, a_o. The latter is one of the basic parameters used in Quantum mechanics. From the BSM point of view, however, the physical meaning of this parameter appears different.

According to BSM concept, the well known parameter a_0 used as a radius in the Bohr model, appears defined only by the

APPENDIX 2

quantum motion conditions of the electron moving in a closed loop with an optimal confined velocity corresponding to an electron energy of 13.6 eV. Then the main characteristic parameter of the quantum loop is not its shape, but its length.

The identity of Eq. (17) and (18) also indicates that **the signature of the fine structure constant is embedded in the quantum loop.**

Now we may use the new obtained meaning about the quantum loop associated with the Bohr orbit, and more specifically the orbital length $2\pi a_0$. For a motion with an optimal confined velocity, the number of electron turns in the quantum orbit is equal to the orbital length divided by the helix step (s_e).

$$\frac{2\pi a_0}{s_e} = \frac{\lambda_c}{\alpha s_e} = 18778.365 \quad \text{turns} \quad (19)$$

Let find at what number of complete orbital cycles (for orbit length of $2\pi a_0$) the phase repetition of the first and second proper frequencies of the electron is satisfied (in other words the smallest number of orbital cycles containing whole number of two frequency cycles). The analysis of the confined motion of the electron in Chapter 3 and 4 of BSM indicates that its secondary proper frequency is three times higher than the first one (the first one is equal to the Compton frequency). Equation (19) shows that the residual number of first proper frequency cycles is close to 1/3. If assuming that it is exactly 1/3 (due to a not very accurate determination of the involved physical parameters), then the condition for phase repetition of both frequency cycles will be met for three orbital cycles. The whole number of turns then should be $3\lambda_c/(\alpha s)_e$. Substituting s_e by its expression given by Eq. (7) we get

$$\frac{3(1-\alpha^2)^{1/2}}{\alpha^2} \quad (20)$$

We have ignored so far the relativistic correction, but for accurate estimation it should be taken into account. The relativistic gamma factor for the electron velocity of $V_{ax} = \alpha c$ is $\gamma = (1-\alpha^2)^{-1/2}$. Multiplying the above expression by the obtained gamma factor we get.

$$3/\alpha^2 = \text{integer} \quad (21)$$

APPENDIX 2

The validity of Eq. (20) and (21) could be tested by the following simple procedure: calculating these expressions by using the best experimental value of α, rounding the result to the closer integer (satisfying the condition for two consecutive phase repetitions) and recalculating the corresponding value of α. The rounded integer (a whole number of turns) could be correct only if the recalculated value is in the range of the accuracy of the experimentally determined α. Let using the recommended value of experimentally measured α according to CODATA 98.

$$\alpha = 7.297352533(27) \times 10^{-3} \quad (\text{CODATA 98})^{16} \quad (22)$$

where, the uncertainty error is denoted by the digits in the brackets.

The calculated values of α from Eq. (20) and (21) exceeds quite a bit the uncertainty value of experimentally determined α given by Eq. (22). Consequently, the condition for phase repetitions of the two proper frequencies is not fulfilled for three orbital cycles with total trace length of $3 \times 2\pi a_0$. Therefore, we may search for the next smallest number of orbital cycles in which the phase repetition conditions are satisfied. It stands to reason that the approximate value of the orbital cycles could be about 137 ($1/\alpha$). Then if not considering relativistic correction, the corresponding number of electron turns is $(1-\alpha^2)/\alpha^3$. When applying a relativistic correction (multiplying by the estimated above gamma factor for the kinetic energy of 13.6 eV) the number of the electron turns becomes $1/\alpha^3$. The phase repetition conditions will be satisfied if this number is integer: $1/\alpha^3 = \text{int} \, eger$

Substituting α by its value from CODATA 98 (Eq. (22)) we get $1/\alpha^3 = 2573380.57$

It is interesting to mention, that the closest integer value of 2573380 is obtained by Michael Wales, using a completely different method for analysis of the electron behaviour (See Michael Wales book "Quantum theory; Alternative perspectives")[17].

We may use one additional consideration, for validation the above obtained number. The number of turns multiplied by the time for one turn (the Compton time) will give the total time on the orbit (or the lifetime of the excited state, according to the Quantum Mechanics terminology). If accepting that the total number of turns are 2573380 then we obtain a lifetime of 2.0827×10^{-14} (s), that appears to be at least two order smaller than the estimated lifetime for some excited states of the atomic hydrogen.

APPENDIX 2

Following the above analysis we may check for phase repetition at $1/\alpha^4$ turns. The participation of α at power of four is in agreement also with the following consideration: In the analysis of the vibrational mode of the molecular hydrogen, an excellent match between the developed model and observed spectra (section 9.7.5 in Chapter 9 of BSM) is obtained if the fine structure constant participates at a power of four. In such case we may accept that the phase repetition conditions is satisfied for a number of turns given by the closest integer in Eq. (23).

$$1/\alpha^4 = \text{integer} \qquad (23)$$

Using the CODATA value of α we obtain $1/\alpha^4 = 352645779.39$. Rounding to the closest integer we obtain an expression for the theoretical value of α (if its experimental estimation is accurate enough).

$$\alpha = (352645779)^{-1/4} = 7.2973525298 \times 10^{-3} \qquad (24)$$

The small difference of the theoretically obtained value of α from the experimental one could be caused by an experimental error. One of the methods for accurate experimental estimation of α is based on the measurement of the Josephson constant, K_J. Its connection to α is given by the expression

$$K_J = \frac{2}{c}\left(\frac{2\alpha}{\mu_0 m_e \lambda_c}\right)^{1/2} \qquad (25)$$

where: μ_0 - is the permeability of vacuum, m_e - is the electron mass, c - is the light velocity, λ_c - is the Compton wavelength.

The accuracy of α according to this method depends mostly on the accuracy of the Josephson constant measurement, because all other parameters are accurately known. The recommended value for this constant according to CODATA 98 is $K_J = 483597.898(19) \times 10^9$ (Hz/V). If replacing α in Eq. (25) with the value given by Eq. (22) we will obtain the value of K_J that is in the uncertainty range given by the CODATA 98.

The conclusion that the orbital time duration may depends only on α is reinforced also by the consideration that the Compton wavelength, λ_c, was initially involved in the analysis (Eq. (15), (17), (19)), but it disappeared in the derived Eq. (23). Consequently, the phase repetition condition is satisfied not only for the two proper frequencies of the electron, but also for the SPM frequency of the

APPENDIX 2

CL nodes included in the quantum orbit (λ_c is the propagated with a speed of light phase of the SPM vector for one SPM cycle of the CL node (SPM frequency = Compton frequency)).

Table II shows the quantum motion parameters of the electron in a quantum loop for velocities corresponding to different subharmonic numbers.

n – is the subharmonic number, E - is the electron energy, V_{ax} - is the axial velocity, V_t - is the tangential velocity of the rotating electron structure, r_{mb} - is the equivalent magnetic radius of the electron limited by the speed of light modulation of the CL nodes from the rotating electron structure, c - is a light velocity, R_c - is the Compton radius, a_o - is the Bohr radius, l_{ql} - is the trace length for a motion in closed loop (single quantum loop), L_q - is the length size of a quantum loop if its shape is a Hippoped curve with a parameter $a = \sqrt{3}$ (close to the shape of digit 8).

Table II. Quantum motion parameters of the electron in a quantum loop

n	E (eV)	V_{ax}	V_t	r_{mb}	l_{ql}	L_q (Å)
1	13.6	αc	c	$\sim R_c$	$2\pi a_0$	1.3626
2	3.4	$\alpha c/2$	$c/2$	$2R_c$	$2\pi a_0/2$	0.6813
3	1.51	$\alpha c/3$	$c/3$	$3R_c$	$2\pi a_0/3$	0.4542
4	0.85	$\alpha c/4$	$c/4$	$4R_c$	$2\pi a_0/4$	0.3406
5	0.544	$\alpha c/5$	$c/5$	$5R_c$	$2\pi a_0/5$	0.2725

The introduced parameter **subharmonic number** shows the rotational rate of the whole electron structure.

7. Quantum orbits.

It is apparent from the provided analysis that a stable quantum loop is defined by the repeatable motion of oscillating electron. The shape of such loop, however, is determined by external conditions. Such conditions may exist in the following two cases:

APPENDIX 2

- a quantum loop obtained between particle with equal but opposite charges and same mass, as in the case of positronium (see Chapter 3 of BSM)

- a quantum loop obtained between opposite charged particles but with different masses (a hydrogen atom as a most simple case and other atoms and ions as more complex cases).

In both options the quantum loops are repeatable and we may call them **quantum orbits**. A single quantum orbit could contain one or few serially connected quantum loops (in both cases the condition for phases repetition is preserved). It is obvious that the shape of the quantum orbit is defined by the proximity field configuration of the proton (or protons). The vacuum space concept of BSM allows unveiling not only the electron structure but also the physical shape of the proton with its proximity electrical field (chapters 6 and 7 of BSM). The shape of any possible quantum orbit is strictly defined by the geometrical parameters of the proton.

Let considering now the induced magnetic field of the electron motion in a quantum orbit by using the electron magnetic radius. The magnetic radius of the electron moving with different subharmonic number n is analysed in section 3.1, Chapter 3 of BSM. Its value for $n = 1$ (a kinetic energy of 13.6 eV) matches the estimated magnetic radius corresponding to the magnetic moment of the electron. For larger numbers (decreased electron energy), however, the magnetic radius shows an increase. The physical explanation by BSM is that at decreased rate of the electron rotation its IG field of the twisted internal RL structure is able to modulate the surrounding CL space up to a larger radius until the rotating modulation of the circumference reaches the speed of light. Keeping in mind that the circumference of the electron is equal to the Compton wavelength (with a first order approximation) the circumference length of the boundary (defined by the rotation rate) should be a whole number of Compton wavelengths. Then the integer number of the Compton wavelengths corresponds to integer subharmonic number. In such case, the orbiting electron with optimal or sub-optimal velocity could not cause external magnetic field beyond some distance from the nucleus. This provides boundary conditions for the atoms, if accepting that in any quantum orbit the electron is moving with optimal or sub-optimal confined velocity (integer sub-harmonic number). Here we must open a bracket that the higher energy levels in heavier elements come not from a larger electron velocity but from the shrunk CL space affected by the accumulated protons and neutrons. Such CL space domain is pumped to larger energy levels in comparison to the CL space surrounding the hydrogen atom.

The existence of the IG law changes significantly the picture of the orbiting electron in a proximity field of the proton. In Chapter 7 of BSM an analysis of Balmer model of Hydrogen atom is developed based on the BSM concept of the electron and proton and the IG law influence on the orbital electron motion in the proximity to the proton. It appears that the limiting orbit has a length of $2\pi a_0$ while all other quantum orbits are inferior. This conclusion is valid not only for the Balmer series in Hydrogen but also for all possible quantum orbits in different atoms, if they are able to provide line spectra. Therefore, the obtained physical model of Hydrogen puts a light for solving **the boundary conditions problem of the electron orbits in the atoms.**

8. Time duration for a stable orbit (lifetime of excited state).

The following analysis could be valid only for the hydrogen, where the influence of the proton mass on the surrounding CL space appears to be negligible.

Keeping in mind the screw-like confined motion, the axial and tangential velocities will be inverse proportional to the subharmonic number. Then the condition for phase repetitions for a motion with a subharmonic number n will be satisfied for n times smaller number of electron turns, or the quantum orbit will be n times smaller. It is reasonable to consider that the first and second proper frequencies of the electron are stable and not dependant on the subharmonic numbers. Then for estimation of the time duration of the orbit (the lifetime of excited state) it is more convenient to use the number of the cycles of the first proper frequency of the electron. It is equal to the number of electron turns for $n = 1$. In such way we arrive to the conclusion:

(a) If conditions for stable quantum orbit are defined only by the phase repetition conditions and the whole number of Compton wavelengths, the time duration (lifetime) of the orbiting electron does not depend on the subharmonic number of its motion.

(b) If (a) is valid, the lifetime of the excited state will be equal to the product of the total number of the first proper frequency electron cycles (according to Eq. (23)) and the

APPENDIX 2

Compton time (the time for one electron cycle with the first proper frequency)

According to condition (b) the theoretical lifetime for an excited state of hydrogen is

$$\tau = t_c/\alpha^4 = \lambda_c/(c\alpha^4) = 2.85407 \times 10^{-12} \text{ (s)} \qquad (26)$$

where: t_c - is the Compton time.

Note: The obtained Eq. (26) does not take into account the possible modification of the surrounding space in a close proximity to the proton. Such modification (a slight shrinkage, or a space curvature) may cause aliasing for the phase repetition conditions due to affected SPM frequency and Compton wavelength, while the first and second proper frequencies of the electron are obviously stable. For heavier atoms such modification may appear much stronger. For elements with more than one electron the mutual orbital interactions also may lead to increase of the real lifetime.

9. Conclusions and comments

According to the BSM hypothesis, the physical model of the electron possesses a structure built by sub-elementary particles, which are also involved in the underlying hypothetical structure of the space (the physical vacuum). The suggested electron model with a signature of anomalous magnetic moment exhibits rich oscillation and interaction behaviour in such space. Two fundamental physical constants as the fine structure constant and the Compton frequency (or wavelength) appear embedded in the electron structure and its dynamical behaviour. The analysis leads to the conclusion that the Compton frequency, v_c, expresses simultaneously two different features: the SPM frequency of the CL node and the first proper frequency of the oscillating electron. At the same time, the Compton wavelength, λ_c, expresses the length of the phase propagation of the SPM vector with a light velocity for one cycle of the SPM frequency of the CL node. This is in agreement with the relation $\lambda_c = c/v_c$. More details about the use of the suggested electron structure for unveiling the meaning of different physical constants are provided in the BSM hypothesis[13]. Further analysis, presented in BSM, leads to derivation of a hydrogen model possessing boundary conditions for the electronic orbits, while exhibiting the same energy levels like the Bohr atomic model. The obtained model of

APPENDIX 2

the hydrogen further served as a base for the suggested spatial arrangement of the protons and neutrons in the atomic nuclei[15].

10. Acknowledgements

I wish to express my gratitude to Mark Porringa for the useful comments and discussions related to the BSM hypothesis and particularly this monograph. Special appreciations and thanks are extended to acad. Prof. Dr. Asparuh Petrakiev of the Burgas University, Bulgaria, for the organized workshop in August 2003, Varvara, Bulgaria and to Angel Manev of the STIL at the Bulgarian Academy of Sciences for the useful discussions.

References (for this article only)

1. L. Ferrarese, D. Merrit, A fundamental relation between supermassive black holes and their host galaxies, http://arxiv.org/abs/astro-ph No. 0006053 v. **2**, 9 Aug 2000
2. D. F. Roscoe, An analysis of 900 optical rotation curves: Dark matter in a corner?, Phahama - journal of physics, Indian Academy of Sciences, Vol. **53**, No 6, Dec 1999, p. 1033-1037
3. T. H. Boyer, The Classical Vacuum, Scientific American, Aug. 1985, p.70-78.
4. H. E. Puthoff, Gravity as a zero-point-fluctuation force, Phys. Rev. A, vol. 39, no 5, 2333-2342, (1989)
5. H. E. Puthoff, Polarizable-Vacuum (PV) Approach to General Relativity, Foundations of Physics, V. 32, No. 6, 927-943 (2002)
6. H. E. Puthoff, Can the Vacuum be Engineered for Spaceflight applications, NASA Breakthrough Propulsion Physics, conference at Lewis Res. Center, (1977)
7. H. E. Puthoff, S. Tittle, M. Ibison, Engineering the Zero-Point Field and Polarizable Vacuum for Interstellar Flight, First International Workshop in Field Propulsion, Univ. of Sussex, Brighton, UK, Jan 2001, http://www.nidsci.org/article3.html
8. B. Haisch, A. Rueda and H. E. Puthoff, Inertia as a Zero-point field lorenz force, Phys. Rev. A, **49**, 678 (1994). See also Science 263, 612 (1994)
9. F. M. Meno, A Planck-length Atomic Kinetic Model of Physical Reality, Physics Essays, **4**, p.94, (1991)
10. M. L. Gershteyn, L. Gershteyn, A. Gershteyn, O. Karagioz, Experimental evidence that the gravitational constant varies with orientation, (2002), http://arxiv.org/abs/physics/0202058
11. M. Ibison, H. E. Puthoff and S. R. Little. The Speed of Gravity Revisited, posted to LANL archives, http://xxx.lanl.gov/abs/physics/9910050

APPENDIX 2

12. S. Sarg, New approach for building of unified theory about the Universe and some results, http://lanl.arxiv.org/abs/physics/0205052
13. S. Sarg, "Basic Structures of Matter", monograph, (2001), http://www.helical-structures.org
  also in National Library of Canada, (2002)
    www.nlc-bnc.ca/amicus/index-e.html (AMICUS No. 27105955) (first edition)
14. S. Sarg, Brief introduction to the Basic Structures of Matter Theory and derived atomic models, Journal of Theoretics, (2003),
    www.journaloftheoretics.com/Links/Papers/Sarg.pdf
15. S. Sarg, Atlas of atomic nuclear structures according to the Basic Structures of Matter theory
    www.journaloftheoretics.com/Links?papers/Sarg2.pdf
    also in Natianl Library of Canada, (2002)
    www.nlc-bnc.ca/amicus/index-e.html
    (AMICUS No. 27106037)
16. P. J. Mohr and B. N. Taylor, CODATA recommended values of the fundamental constants: 1998, Rev. Mod. Phys, **72**, 351-495, (2000)
17. M. Wales, Quantum Theory; Alternative Perspectives, Shields Books, www.fervor.demon.co.uk

About the Author

Stoyan Sarg - Sargoytchev is a Bulgarian-born Canadian. He holds an engineering diploma and a PhD in physics in the field of space research. From 1976 to 1990 he was actively involved in projects sponsored by the international program Intercosmos, coordinated by the former Soviet Union. During this period he also participated in a collaborative program with the European Space Agency. For his pioneering work in space research, Dr. Sargoytchev has been awarded medals from the Bulgarian government and from the Intercosmos organization.

In 1990, Dr. Sargoytchev was invited as a visiting scientist to Cornel University and worked in Arecibo Observatory, P.R. on a Lidar project funded by the NSF (USA). This was the place where the SETI (Search for Extraterrestrial Intelligence) program was partly operated before 1986 using the world largest radiotelescope – radar. From September 1991 to June 1993 he was involved in building the Lidar system at the University of Western Ontario, Canada. In 1993 he was employed as a project scientist in the Institute for Space and Terrestrial Science (later CRESTech) working on space projects coordinated by the Canadian Space Agency. Since 2002 he was with York University, Toronto, Canada.

Working on diversified projects, Dr. Sargoytchev obtained valuable experience and knowledge in different fields of Space Engineering and Physics. During his 30 years working in academic institutions, he participated in many conferences and seminars, while understanding that contemporary physics is plagued by unsolved problems and contradictions. He paid particular attention to unsolved mysteries – physical phenomena, for which contemporary physics still does not offer any explanation. After studying extensively the history of physics in the past three centuries, Dr. Sargoytchev realized that the origin of all problems is in the adopted concept about space. Focusing on this issue he arrived at a new idea about space, time and matter using an original classical approach and physical models of hierarchical material structure converging at the level of the Planck scale (Planck's length: 1.616×10^{-35} (m), Planck's time: 5.39×10^{-44} (s)). After a few years of intensive but rewarding work he was able to develop and publish in 2001 his treatise, called Basic Structures of Matter - Supergravitation Unified Theory (BSM-SG). While the work was based on a challenging revolutionary new idea, it did not match any established program, so he published it as a monograph under the name Stoyan Sarg. While having over 70 scientific publications (including eight patents) in the field of space research Dr. Stoyan Sarg Sargoytchev considered his theoretical work as a major achievement in his life. He was convinced that he had found a solution to a fundamental problem in Physics in a way that has been considered impossible in the past 100 years: building a successful unified theory in a real 3-dimensional space with revealed relations between the gravitational, electric and magnetic fields. The monograph "Basic Structures of Matter" has two electronic editions archived in the National Library in Canada (2002) and a few publications in scientific journals. In 2006 he published the whole theory as a book entitled "Basic Structures of Matter – Supergravitation Unified Theory". The book includes a previous edited electronic version and a new Chapter 13 summarizing the theoretical results and predicted applications, particularly in the field of energy and gravitation. A short popular version of his theory was previously published in 2004 as a book titled "Beyond the Visible Universe". A reviewing committee from the Canadian Association of Physicists published a review of his two books in Physics in Canada Journal vol. 62, No 4, 207-207, (2006). Related articles are published in peer reviewed journals, on-line journals, and reports in international scientific conferences. The original work of Dr. Sarg is indexed by multiple citations on the Internet.

Amongst the conclusions for potential applications predicted by BSM-SG theory, the most important are the following:

(1) Unveiling of hidden space energy of non EM type – a primary source of the nuclear energy;

(2) A possibility to control the gravitational and inertial mass of a material object;

(3) A possibility for supercommunication by waves different from EM waves.

The first conclusion provides a new vision about accessing the energy existing in space, known as energy from physical vacuum. The second conclusion predicts development of a completely new propulsion mechanism suitable for distant space travels. The third conclusion envisions the possibility for a new type of communications with superluminal velocity.

After development of the theory, the author focused on the second conclusion – the possibility for obtaining a propulsion effect by unidirectional change of the gravitational and inertial mass. The physical models of BSM-SG appeared quite useful for providing of successful laboratory experiments in a comparatively short time.

The experiments were reported at the Annual meetings of the Society for Scientific Explorations in 2007 and 2008. The theoretical models, an invention about a new propulsion method and the supported experiments are described in the book "Field Propulsion by Control of Gravity". The author hopes that his discovery will contribute for the future advance in the interplanetary and deep space travels.

For more information, search the Internet by keywords: S. Sarg; Stoyan Sarg; Basic Structures of Matter; Super Gravitation Unified Theory, Cosmic Lattice, SARG effect, gravito-inertial propulsion effect.

Theory and Experiments

References

1. S. Sarg, New approach for building of unified theory, http://lanl.arxiv.org/abs/physics/0205052 (May 2002)
2. S. Sarg, A Physical Model of the Electron according to the Basic Structures of Matter Hypothesis, Physics Essays, vol. 16 No. 2, 180-195, (2003); http://www.physicsessays.com
3. S. Sarg, Basic Structures of Matter Hypotheses based on an Alternative Concept of the Physical Vacuum, World year of Physics,- 2005 Physics for the Third Millennium: II, Conference organized by NASA, 5-7 Apr 2005, Huntsville, Alabama, USA
4. Stoyan Sarg, *Basic Structures of Matter –Supergravitation Unified Theory*, Trafford Publishing, 2006, ISBN 1412083877 (books review in "**Physics in Canada**,", v. 62, No. 4, July/Aug, 2006)
5. S. Sarg, BSM - Supergravitation unified theory based on an alternative concept of the physical vacuum, IX Int. Sci. Conference "Matter, Energy, Gravitation", 7-11 Aug 2006, St. Petersburg, Russia
6. S. Sarg, Brief Introduction to the Basic Structures of Matter Theory and Derived Atomic Models
   www.journaloftheoretics.com/Links/Papers/Sarg.pdf
7. Atlas of Atomic Nuclear Structures,
   www.nlc-bnc.ca/amicus/index-e.html
   (AMICUS No. 27105955), (2002)
8. S. Sarg, BSM Supergravitation Unified Theory, based on a new concept of the physical vacuum. Major Predictions. International Conference on Future Energy (COFE-2006) 22-24 Sep 2006, Washington DC
9. S. Sarg, Unified Theory Based on an Alternative Space Concept: Predictions for a New Propulsion Mechanism and Experimental Results, 26 Annual SSE Meeting, May 30 – June 2, 2007, East Lansing, Michigan
10. S. Sarg, Gravito-inertial Propulsion Effect Predicted by the BSM - Supergravitation Unified Theory, 27 Annual SSE Meeting, June 25-28, 2008, Boulder, CO, USA
11. S. Sarg, Method and Apparatus for Spacecraft Propulsion with a Field Shield Protection, Patent application in Canada (File No. 2,638,667 from 26 Aug 2008).
12. J. C. Maxwell, *A Treatise on Electricity and Magnetism* vol. II

13. A. Einstein, *Essays in Science*, p. 19; Philosophical Library, NY, 1934.
14. A. Einstein, *Sidelights on Relativity, 1922*
15. Alternative Cosmology Group http://www.cosmology.info/
16. Michelson, A. A. and Morley, E. W. "On the Relative Motion of the Earth and the Luminiferous Aether." *Philos. Mag.* **24**, 449-463, 1887.
17. R. R. Hatch, Those scandalous clocks, GPS Solutions, 8:67–73, (2004)
18. S. Marinov, Measurement of the Laboratory's Absolute velocity, General Relativity and Gravitation, **12**, 57-66, (1980)
19. Marinov, The interrupted "rotating disc" experiment, J. Phys A: Math. Gen, **16**, 1885-1888, (1983)
20. G. Sagnac, Regarding the Proof of the Existence of a Luminifeorous Ether Using a Rotating Interferometer Experiment, (English translation by W. Lonc, The Abraham Zelmanov Journal, 77-8-, (2008).
21. K. J. van Vlaenderen and A. Waser, Generalization of classical electrodynamics to admit a scalar field and longitudinal waves, Hadronic Journal, **24**, 609-628, (2001).
22. K P. Butusov, Longitudinal waves in vacuum: Creation and Research, New Energy Technologies, Sep- Oct 2001, pp. 46-47.
23. Quian-shen Wang et al, Presize measurement of gravity variation during a total solar eclipse, Physical review D, 62, 041101(R), (2000).
24. H. E. Puthoff, Gravity as a zero-point fluctuation force, Physical Review A, **39**, No. 5, 2333-2342, (1989) P. A.
25. И. И. Добромыслов, Свойство спиральной поляризации гравитационных волн, Прикладная физика 4-2003, УДК 530.12:531.51
26. G. Hodowanec, "OP-AMP Circuit Detects Gravity Signal", Radio-Electronics, April 1986
27. G. Hodowanec, archived material in rexresearch www.rexresearch.com
28. R. Monastersky, Science news, 25 July, 1998.
29. N. A. Kozirev, *Selected Works*, (Russian Lang.). 1991.
30. Paul Hill, *Unconventional Flying Objects a scientific analysis*, ISBN 1-57174-027-9, Hampton Roads publishing Company Inc, Charlottesville, VA,

31. Peter Sturrock, *The UFO Enigma, a new review of the physical evidence*, ISBN: 0-446-5265-0, Published by Time Warner Company, (1999).
32. Society for Scientific Exploration www.scientificexploration.org
33. Integrity Research Institute www.integrityresearchinstitute.org
34. Center for UFO Study www.cufos.org
35. S. T. Friedman www.stantonfriedman.com www.frontierscience.us/article128.html
36. W. Treurniet www.treurniet.ca/UFO
37. J. Hutchison, http://www.hutchisoneffect.ca/
38. J. Hutchison, The Hutchison Effect Apparatus, Electric Spacecraft Journal, Issue 9, 14-21, (1993).
39. Lord Kelvin, On the Generation of Longitudinal Waves in Ether, Proceedings of the Royal Society of London, **59**, 270-273, (1895-1896)
40. C. Monstein and J. P. Wesley, Observation of scalar longitudinal electromagnetic waves, Electrophysics Letters, **59**, (4), 514-520 (2002)
41. Massines et al. Experimental and theoretical study of a glow discharge at atmospheric pressure controlled by dielectric barrier, J. Appl. Phys. 83, 2950 (1998)
42. N. Tesla, Experiments with Alternative Currents of High Potential and High Frequency, report delivered before the Institution of Electrical Engineers, London, February 1892, Twenty First Century Books, Breckenridge, CO, USA.
43. D. Mugnai, A. Ranfagni, and R. Ruggeri, Observation of Superluminal Behavior in Wave Propagation, Phys. Rev. Lett., v. 84, No 21, 4830-4833, (2000).
44. T. W. Barrett, Tesla's nonlinear oscillator-shuttle-circuit (OSC) theory, Annales de la Fondation Louis de Broglie, V. 16, No 1, 23-41, (1991)
45. S. Sarg, Sarg Antigravity Exp 1, May 16, 2007, (youtube.com)
46. S. Sarg, Sarg Antigravity Experiment 2 (in dark), October 5, 2007, (youtube.com)
47. S. Sarg, Sarg Antigravity Experiment 2 (in light), October 5, 2007, (youtube.com)

48. S. Sarg, Sarg Propulsion effect, June 22, 2008 (youtube.com)
49. J. R. Roth, D. M. Sherman and S. P. Wilkinson, Electrohydrodynamic flow control with a glow-discharge surface plasma, AIAA Journal, 38, No 7, (2000)
50. http://en.wikipedia.org/wiki/Biefeld-Brown_effect
51. www.helical-structures.org
52. J. N. Munday, F. Capasso, A. Parsegian, measured long-range repulsive Casimir-Lifshitz forces, Nature, **457**, 170-173, (2009).
53. C. Rutkowski & G. Dittman, *The Canadian UFO Report*, Dundurn Press, Tonawanda, NY
54. J. R. Roth, Industrial Plasma Engineering, Volume 1 &Volume II – Application to Non-thermal Plasma Processing, Institute of Physics Publishing, Bristol and Philadelphia.
55. W. B. Smith , www.rexresearch.com/smith/newsci.htm
56. T. Good, Alien Base, Avon Books, Inc, N. Y. , 1999.
57. Angerth, B., Block, L., Fahleson, U. V., and Soop, K.: 1962, *Nucl. Fusion Suppl.* Pt. 1, 39.
58. G. Himmel and A. Piel, Velocity limitation in a rotating plasma device of homopolar type, J. Phys. D: Applied Physics Letters, 6, L108, (1973).
59. R. Haines www.nicap.org/bios/haines.htm
60. R. Dean http://www.beyondzebra.com/bobdean.shtml
61. W. B. Smith, http://sciencevista.com/?p=26
62. G. Hathaway, Mindbending: The Hutchison Files: 1981 to 1995 (ISBN 978-0-9813785-0-3)
63. J. L. Naudin, http://jnaudin.free.fr

Made in the USA
Middletown, DE
29 September 2016